MECHANICS OF DEFORMA

Graduate Student Series in Physics
Other books in the series

Colour Centres and Imperfections in Insulators and Semiconductors
P D Townsend and J C Kelly

The Superconducting State
A D C Grassie

Refrigeration and Thermometry below One Kelvin
D S Betts

Plasma Physics
E W Laing

Weak Interactions
D Bailin

Gauge Theories in Particle Physics
I J R Aitchison and A J G Hey

Collective Effects in Solids and Liquids
N H March and M Parrinello

Hadron Interactions
P D B Collins and A D Martin

Superfluidity and Superconductivity
D R Tilley and J Tilley

GRADUATE STUDENT SERIES IN PHYSICS

Series Editor: Professor Douglas F Brewer M.A., D.Phil.
Professor of Experimental Physics, University of Sussex

MECHANICS OF DEFORMABLE MEDIA

A B BHATIA
(deceased)
Formerly of University of Alberta, Canada

R N SINGH
University of Alberta, Canada
and
University of Bhagalpur, India

ADAM HILGER, BRISTOL AND BOSTON
Published in association with the University of Sussex Press

British Library Cataloguing in Publication Data

Bhatia, A.B.
Mechanics of deformable media. ——
(Graduate student series in physics)
1. Continuum mechanics
I. Title II. Singh, R.N. III. Series
531 QA808.2

ISBN 0-85274-499-4
ISBN 0-85274-500-1 Pbk

Published by IOP Publishing Ltd under the Adam Hilger imprint
Techno House, Redcliffe Way, Bristol, BS1 6NX, England
PO Box 230, Accord MA 02018, USA
in association with the University of Sussex Press

Phototypeset by Mid-County Press, 2a Merivale Road, London SW15 2NW

Printed in Great Britain by J W Arrowsmith Ltd, Bristol

PREFACE

'Mechanics of Deformable Media' has been written with the hope of providing the necessary background to the physics of the continuum theory of condensed systems. It is also our hope that such a comprehensive course on solids and fluids, as here, should be useful to students of physics, material science or geophysics.

This book has emerged from the lecture notes given by one of us (ABB) for a number of years as a one-term course at the Department of Physics, University of Alberta. In fact, the idea to write the book arose out of suggestions from several students and colleagues that it would be useful. As may be seen from the contents, no single book at present covers all this material—and yet the whole topic has a unity of its own. No claims should be made regarding the completeness of the subject matter of the individual chapters; nonetheless, it lays a foundation stone for further specialised training.

It is most regrettable to mention the sudden and sad demise of Professor A B Bhatia while the writing of this text was in an advanced stage. Obviously the text is deprived of his finishing touch and the responsibilities for errors and omissions solely rest with the second author.

The authors take great pleasure in acknowledging the direct or indirect help received from the members of staff of this department. Our special thanks are, however, due to Professor M Razavy and Professor N H March for reading the manuscript critically and for many valuable suggestions. One of us (RNS) expresses his gratitude to the Natural Sciences and Engineering Research Council of Canada for partial support and the Department of Physics, University of Alberta for hospitality. We are also thankful to Mrs Mary Yiu for typing the manuscript with efficiency.

A B Bhatia
R N Singh

CONTENTS

1

OUTLINE

1.1 Preliminary remarks

When we get away from the realm of rigid bodies (idealised case of true state of affairs) which cannot change their shape and size—we have qualitatively two different types of deformable media. First, fluids which at rest cannot support shearing forces, and hence the physical variables like pressure and density (or volume) are sufficient for their description. They do not possess any definite shape, but take the shape of the vessel they are contained in. Solids, on the other hand, can support shearing stresses and can take arbitrary shapes, as for example, a given substance such as aluminium in the form of a sheet, a bar or a ball.

Whenever a material medium is disturbed (for instance, stretched, twisted or deformed in any other way) from its original condition, each element of the matter in it is, in general, displaced. The problem of specifying these deformations† along with their influence on the physical properties forms the core material of the book. To describe the deformations in solids one needs more variables than just the pressure and density. Also, for fluids in motion or in liquid crystals there are frictional and shearing forces which come into play and so the formalism developed for the case of the solid is also useful for fluids. Hence, despite their apparent complexity, we start with solids. We shall mostly be discussing solids in the first half of the book and fluids in the second half.

1.2 Plan of the book

In Chapter 2 we review the chief characteristics of scalars, vectors and tensors with due emphasis on how they transform under orthogonal transformations of axes. Chapter 3 has been utilised to analyse the general deformation in a solid where deformation, strain and rotational tensors are introduced and are amply discussed. In Chapter 4, we first introduce stress tensors and derive the equilibrium equation. Furthermore, an expression for the work done in deforming the system infinitesimally is derived. This is then used to define the strain-energy function and thence introduce the generalised Hooke's law. Stress–strain relations and elastic constants (Lame's constants) for an elastically isotropic body using the tensor invariants, introduced in Chapter 2,

† In this book we are concerned with small elastic deformations only.

are derived next. This is followed by a discussion of the appropriate elastic constants for various homogeneous deformations. In Chapter 5 we specialise the equilibrium equations of Chapter 4 to an elastically isotropic body with an application to calculate the stress distribution in a long cylindrical pipe.

Introducing the concept of the local and material derivative of a function, the general equations of motion are first set up in Chapter 6 in terms of stress derivatives valid for any material—solid or fluid. They are then used to discuss the wave propagation in an infinite elastically isotropic material, showing that only two types of wave disturbances namely irrotational (dilatational) and shear (isovoluminous) are possible. The reflection and the refraction of plane waves at the interface of two materials is discussed and the phenomena of mode conversion occurring here and in scattering by grains is briefly described. In Chapter 7 we make use of the fact that elastic constants are tensors of rank four, to demonstrate how the number of elastic constants reduce depending upon the crystal symmetry and isotropic material. These general tensor relations may be used to derive average values appropriate to a polycrystalline material which has been briefly described.

Chapter 8 is exclusively devoted to the thermodynamics of solids. We first write down the combined statement of the first and second laws of thermodynamics in terms of a full set of stress and strain variables, and we then derive the various useful relations. The special case of cubic materials is discussed and the (thermoelastic) damping of plane waves and other vibrations due to the flow of heat is briefly introduced.

Fluids at rest, by definition, cannot sustain shear stresses and, therefore, the general form of the equilibrium equations derived in Chapter 4 greatly simplifies in what is shown in Chapter 9. This is then used to discuss such problems as the pressure of a fluid at different points in a potential field, the free surface of a liquid in a rotating drum, centrifuge and barometric formula for isothermal and adiabatic atmospheres. In Chapter 10 we undertake the flow problems in perfect fluids (inviscid or frictionless) and introduce the basic equation of continuity, Euler's equation of motion and Bernoulli's equation along with some simple applications. The problems of potential flow in three dimensions for a perfect incompressible fluid are treated in Chapter 11, and in two dimensions are discussed in Chapter 12 using the complex-variable method.

Fluids with viscosity are considered in Chapter 13. The viscosity stress tensor, which is immediately written down using the symmetry consideration given in Chapter 4, is then utilised to obtain the Navier–Stokes (NS) equation. Energy dissipation due to viscosity is briefly introduced. Chapter 14 deals with the propagation and absorption of sound and viscous waves in a real fluid. The coupling of the NS equations with the heat diffusion equation, and sound absorption due to viscosity and thermal conductivity, are discussed. Finally in Chapter 15 we consider briefly some simple flow problems (such as laminar flow, Reynold's number, Prandtl's boundary layer) in a real fluid.

Each Chapter is followed by a few problems. Most of these have evolved in the process of teaching or in preparing the manuscript. A few problems, however, have their roots in various existing text books and due to their importance they find their place here.

2

SCALARS, VECTORS AND TENSORS

A physical quantity which requires just one number for its specification is called a scalar. For example, the distance between two points in space or the speed of a particle are scalar quantities. On the other hand, a vector, like the line joining two points in space or force acting on a particle, has both direction and magnitude. A vector in n-dimensional space requires n numbers for its specification. The actual values of these numbers (components) depend on the choice of coordinate axes. The manner in which a scalar, components of a vector and components of a tensor (to be defined presently) transform when one goes from one reference Cartesian coordinate system to another is essential to the understanding of the subject matter of this book. This Chapter reviews these transformation properties.

2.1 Orthogonal transformation of coordinates

2.1.1 Description and notation

We consider three-dimensional space specifically. Let the axes of the two rectangular Cartesian frames of reference, with common origin O, be labelled x_1, x_2, x_3 and x'_1, x'_2, x'_3 as depicted in figure 2.1. The coordinates of a general point P in the two coordinate systems are respectively denoted by (x_1, x_2, x_3) and (x'_1, x'_2, x'_3). Obviously these two sets of numbers are alternate ways of describing where the point P is. Now let all a_{11}, a_{12}, a_{13} denote the cosines of the angles which the x'_1-axis makes, respectively, with the three coordinate axes x_1, x_2, x_3 of the unprimed frame. Then the coordinate x'_1 of P is related to the coordinates x_i $(i = 1, 2, 3)$ by

$$x'_1 = a_{11}x_1 + a_{12}x_2 + a_{13}x_3.$$

Similarly with a_{11}, a_{21}, a_{31} representing the direction cosines of the x_1-axis with respect to the primed axes, one has

$$x_1 = a_{11}x'_1 + a_{21}x'_2 + a_{31}x'_3.$$

The equations for other components follow in a similar fashion and a convenient way of obtaining them is by reading along the rows and columns of table 2.1. In abbreviated form the equations are, $(i, j = 1, 2, 3)$:

$$x'_i = \sum_{j=1}^{3} a_{ij}x_j \equiv a_{ij}x_j \qquad (2.1a)$$

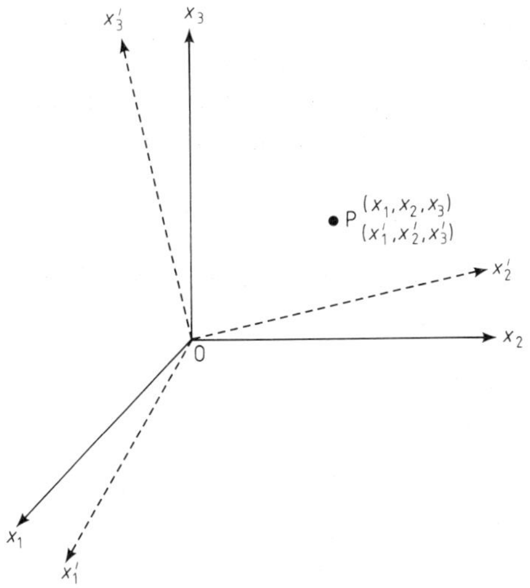

Figure 2.1 Two Cartesian coordinate systems with a common origin.

Table 2.1 Relation between the primed and unprimed orthogonal coordinate axes.

	x_1	x_2	x_3
x_1'	a_{11}	a_{12}	a_{13}
x_2'	a_{21}	a_{22}	a_{23}
x_3'	a_{31}	a_{32}	a_{33}

$$x_i = \sum_{j=1}^{3} a_{ji} x_j' \equiv a_{ji} x_j' \tag{2.1b}$$

where in writing the identities we have followed the Einstein summation convention—whenever two indices are the same a sum over repeated indices is implied.

Remembering that a_{ij} represent direction cosines and that x_1, x_2, x_3 axes are mutually orthogonal to one another, as the primed axes are also, one has

$$a_{il} a_{kl} = \delta_{ik} \tag{2.2a}$$

$$a_{li} a_{lk} = \delta_{ik} \tag{2.2b}$$

where δ_{ik} is the Kronecker delta:

$$\begin{aligned} \delta_{ik} &= 0 \qquad \text{if } i \neq k \\ &= 1 \qquad \text{if } i = k. \end{aligned} \tag{2.3}$$

Note that each of equations (2.2*a*) and (2.2*b*) are six relations; they are, of course, not all independent of one another.

In future, for brevity, we refer to the above transformation of Cartesian coordinate axes and coordinates as simply an orthogonal transformation.

2.1.2 Determinant of $a_{ij}[\equiv det(a_{ij})]$

The transformation law (2.1*a*) can be written as a matrix equation with x'_i and x_i as column matrices, thus

$$\begin{Bmatrix} x'_1 \\ x'_2 \\ x'_3 \end{Bmatrix} = \begin{Bmatrix} a_{11} & a_{12} & a_{13} \\ a_{21} & a_{22} & a_{23} \\ a_{31} & a_{32} & a_{33} \end{Bmatrix} \begin{Bmatrix} x_1 \\ x_2 \\ x_3 \end{Bmatrix} \tag{2.4}$$

or, in abbreviated form:

$$\boldsymbol{x}' = \mathbf{a}\boldsymbol{x}. \tag{2.5a}$$

If $\mathbf{a}^{-1}$ denotes the inverse matrix of $\mathbf{a}$, then from (2.5*a*) one has

$$\boldsymbol{x} = \mathbf{a}^{-1}\boldsymbol{x}'. \tag{2.5b}$$

Comparing this with (2.1*b*) one sees that $\mathbf{a}^{-1}$ is just $\mathbf{a}^{\mathrm{T}}$, the transposed matrix of $\mathbf{a}$. Since $\det \mathbf{a}^{\mathrm{T}} = \det \mathbf{a}$, it follows that

$$\det \mathbf{a}^{\mathrm{T}} = \det \mathbf{a} = \det \mathbf{a}^{-1}. \tag{2.6}$$

Furthermore, one has, by definition

$$\mathbf{a}^{-1}\mathbf{a} = \mathbf{a}\mathbf{a}^{-1} = \mathbf{I} \tag{2.7}$$

where $\mathbf{I}$ is the unit matrix:

$$\mathbf{I} = \begin{Bmatrix} 1 & 0 & 0 \\ 0 & 1 & 0 \\ 0 & 0 & 1 \end{Bmatrix}. \tag{2.8}$$

Now from (2.7)

$$\det \mathbf{a}^{-1}\mathbf{a} = 1 = \det \mathbf{a}^{-1} \det \mathbf{a}$$

and hence using (2.6)

$$\det \mathbf{a} \det \mathbf{a} = 1 \qquad \text{or } \det \mathbf{a} = \pm 1 \tag{2.9}$$

which is the relation we wished to obtain.

The plus sign in equation (2.9) applies when the transformed axes x' are obtained by pure rotation of the old coordinate system. It leaves a right-handed coordinate system[†] right handed, and a left-handed coordinate system

[†] In a right-handed system, the axes are so directed that if the head of a right-handed screw moves from $x_1 \to x_2$, the screw advances along the positive x_3-axis.

left handed. As an example, consider the rotation of a right-handed system through 180° about the x_3-axis (figure 2.2); one has:

$$\mathbf{a} = \begin{Bmatrix} -1 & 0 & 0 \\ 0 & -1 & 0 \\ 0 & 0 & 1 \end{Bmatrix} \tag{2.10}$$

and hence det $\mathbf{a} = +1$.

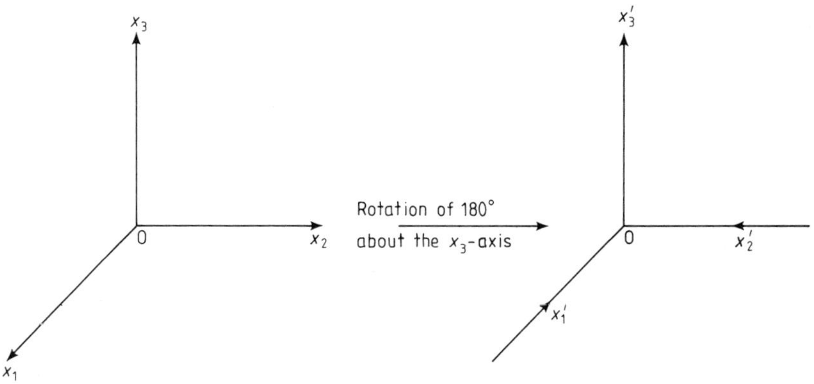

Figure 2.2 The rotation of a right-handed coordinate system through 180° about the x_3-axis.

The negative sign in equation (2.9), on the other hand, signifies that the new axes are obtained by a reflection in a plane or by a rotation which is followed (or preceded) by a reflection in a plane. It changes a right-handed coordinate system into a left-handed system and *vice versa*. Consider, for example, a reflection in the $x_1 - x_2$ plane (figure 2.3). Then

$$\mathbf{a} = \begin{Bmatrix} 1 & 0 & 0 \\ 0 & 1 & 0 \\ 0 & 0 & -1 \end{Bmatrix} \tag{2.11}$$

and hence det $\mathbf{a} = -1$.

We refer, for brevity, to the two cases of positive and negative signs in equation (2.9) as transformation (a) and (b) respectively.

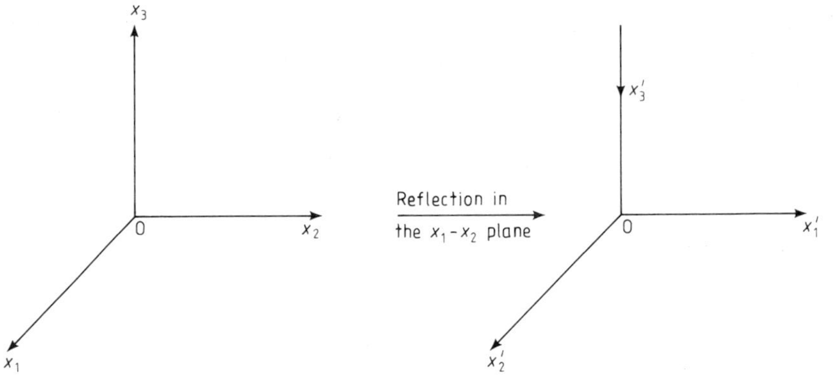

Figure 2.3 The reflection of a right-handed coordinate system in the x_1–x_2 plane. Note that the primed axes system here is left handed.

2.2 Scalars and vectors

2.2.1 Definitions

A scalar quantity possesses just magnitude, like the distance between two points, say O and P. The distance OP remains the same irrespective of how we draw our axes. For orthogonal transformations we have

$$\mathrm{OP}^2 = x_1^2 + x_2^2 + x_3^2 = x_1'^2 + x_2'^2 + x_3'^2. \tag{2.12}$$

Quite generally a physical quantity f is a scalar if it is invariant under orthogonal transformations, i.e., if

$$f(x_1, x_2, x_3) = f(x_1', x_2', x_3') \tag{2.13}$$

it being understood at the cost of repetition that, as in (2.12), the x_i' refer to the same point in space as the x_i. Quantities such as the speed of a particle, and electric and gravitational potentials, are examples of scalars.

A vector has both magnitude and direction. We call a physical quantity, $\boldsymbol{A}$, a vector in three-dimensional space if its components, say A_1, A_2, A_3, transform to A_1', A_2', A_3' like the coordinates x_i under an orthogonal transformation, i.e. (cf equation (2.1))

$$A_i' = a_{ij} A_j \tag{2.14a}$$

$$A_i = a_{ji} A_j'. \tag{2.14b}$$

Clearly if $\boldsymbol{A}$ and $\boldsymbol{B}$ are two vectors any linear combination, $\boldsymbol{C} = a\boldsymbol{A} + b\boldsymbol{B}$, ($a$, b are constants) is also a vector. Physical quantities like the position vector OP, velocity, force, electric field, current density, etc are all vector quantities.

There are some quantities which transform like a vector under rotation (transformation (a) above), but which change sign under reflection (transformation (b)). These are referred to as axial vectors—to distinguish them from the ordinary or polar vectors which transform like (2.14) for both rotations and reflections. Unless specified otherwise a vector in the following will refer to an ordinary vector.

If $\boldsymbol{A}$ and $\boldsymbol{B}$ are (ordinary) vectors, their vector product, $\boldsymbol{C}=\boldsymbol{A}\times\boldsymbol{B}$, is an axial vector and so also is $\boldsymbol{D}=\text{curl}\,\boldsymbol{A}$ (see below). Physical quantities like the moment of a force, vorticity and magnetic field behave like axial vectors.

2.2.2 *More on scalars and vectors*

Some examples relating to the use of transformations which the reader should familiarise himself with are given below.

(i) If f is a scalar then grad f is a vector.

Denoting $\boldsymbol{n}_1, \boldsymbol{n}_2, \boldsymbol{n}_3$ as unit vectors along x_1-, x_2-, x_3-axes respectively, grad f may be written as

$$\text{grad } f \equiv \nabla f = \left(\boldsymbol{n}_1 \frac{\partial}{\partial x_1} + \boldsymbol{n}_2 \frac{\partial}{\partial x_2} + \boldsymbol{n}_3 \frac{\partial}{\partial x_3}\right) f. \tag{2.15}$$

Abbreviating its components $\partial f/\partial x_i$ by B_i $(i=1,2,3)$, we have

$$\frac{\partial f}{\partial x_i'} \equiv B_i' = \frac{\partial f}{\partial x_j}\frac{\partial x_j}{\partial x_i'} = \frac{\partial f}{\partial x_j} a_{ij}$$

on using (2.1). Hence

$$B_i' = a_{ij} B_j. \tag{2.16}$$

Equation (2.16) demonstrates that B_i transform like the components of a vector and hence grad f is a vector.

(ii) The scalar product of two vectors $\boldsymbol{A}$ and $\boldsymbol{B}$ is a scalar.

Remembering (2.14) and (2.2), we have

$$\begin{aligned}\boldsymbol{A}'\cdot\boldsymbol{B}' \equiv A_i'B_i' &= (a_{ij}A_j)(a_{ik}B_k) \\ &= (a_{ij}a_{ik})A_jB_k = A_jB_j = \boldsymbol{A}\cdot\boldsymbol{B}.\end{aligned} \tag{2.17}$$

which proves that $\boldsymbol{A}\cdot\boldsymbol{B}$ is a scalar.

(iii) div $\boldsymbol{A}$ is a scalar.

The divergence of a vector $\boldsymbol{A}$ is defined as

$$\begin{aligned}\text{div}\,\boldsymbol{A} &\equiv (\nabla\cdot\boldsymbol{A}) \\ &= \left(\boldsymbol{n}_1 \frac{\partial}{\partial x_1} + \boldsymbol{n}_2 \frac{\partial}{\partial x_2} + \boldsymbol{n}_3 \frac{\partial}{\partial x_3}\right)\cdot(\boldsymbol{n}_1 A_1 + \boldsymbol{n}_2 A_2 + \boldsymbol{n}_3 A_3) \\ &= \frac{\partial A_1}{\partial x_1} + \frac{\partial A_2}{\partial x_2} + \frac{\partial A_3}{\partial x_3} \equiv \frac{\partial A_i}{\partial x_i}.\end{aligned} \tag{2.18}$$

Consider now

$$\text{div } \boldsymbol{A}' = \frac{\partial A_i'}{\partial x_i'} = \sum_{ik} \frac{\partial}{\partial x_i'} (a_{ik} A_k)$$

$$= \sum_{ijk} \left(\frac{\partial x_j}{\partial x_i'} \right) \frac{\partial}{\partial x_j} (a_{ik} A_k)$$

$$= \sum_{ijk} a_{ij} a_{ik} \frac{\partial A_k}{\partial x_j}.$$

On using (2.2) this becomes

$$\text{div } \boldsymbol{A}' = \text{div } \boldsymbol{A} \tag{2.19}$$

which shows that div $\boldsymbol{A}$ is a scalar quantity.

(iv) The vector product of two vectors is an axial vector.

The vector product of two vectors $\boldsymbol{A}$ and $\boldsymbol{B}$ is defined by

$$\boldsymbol{A} \times \boldsymbol{B} \equiv \boldsymbol{C} = \begin{pmatrix} \boldsymbol{n}_1 & \boldsymbol{n}_2 & \boldsymbol{n}_3 \\ A_1 & A_2 & A_3 \\ B_1 & B_2 & B_3 \end{pmatrix} \tag{2.20}$$

whose components are

$$\begin{aligned} C_1 &= A_2 B_3 - A_3 B_2 \\ C_2 &= A_3 B_1 - A_1 B_3 \\ C_3 &= A_1 B_2 - A_2 B_1. \end{aligned} \tag{2.21}$$

As mentioned earlier $\boldsymbol{C}$ is an axial vector. To verify this consider first the rotation of axes through 180° about the x_3-axis. Referring to equations (2.10) and (2.14), one has

$$\begin{matrix} A_1' \to -A_1 & A_2' \to -A_2 & A_3' \to A_3 \\ B_1' \to -B_1 & B_2' \to -B_2 & B_3' \to B_3 \end{matrix} \tag{2.22a}$$

and hence

$$\begin{aligned} C_1' &= -A_2 B_3 + A_3 B_2 = -C_1 \\ C_2' &= -A_3 B_1 + A_1 B_3 = -C_2 \\ C_3' &= A_1 B_2 - A_2 B_1 = +C_3. \end{aligned} \tag{2.22b}$$

Equations (2.22) illustrate that $\boldsymbol{C} = \boldsymbol{A} \times \boldsymbol{B}$ transforms like an ordinary vector for the rotations. On the other hand, for reflection in the plane x_1–x_2, one has using (2.11) and (2.14)

$$\begin{matrix} A_1' \to A_1 & A_2' \to A_2 & A_3' \to -A_3 \\ B_1' \to B_1 & B_2' \to B_2 & B_3' \to -B_3 \end{matrix} \tag{2.23a}$$

and hence

$$\begin{aligned} C'_1 &= A'_2B'_3 - A'_3B'_2 = -C_1 \\ C'_2 &= A'_3B'_1 - A'_1B'_3 = -C_2 \\ C'_3 &= A'_1B'_2 - A'_2B'_1 = +C_3 \end{aligned} \tag{2.23b}$$

so that here $C'_i = -(a_{ij}C_j)$. We may recall that a quantity represented by a vector product involves the sense of rotation (clockwise or anticlockwise) and its direction is determined by the right-handed screw rule. One has

$$\boldsymbol{A}\times\boldsymbol{B} = -\boldsymbol{B}\times\boldsymbol{A}. \tag{2.24}$$

(v) curl $\boldsymbol{A}$ is an axial vector.

The curl of a vector $\boldsymbol{A}$ is defined as

$$\text{curl } \boldsymbol{A} \equiv \boldsymbol{D} \equiv \nabla\times\boldsymbol{A} = \begin{pmatrix} \boldsymbol{n}_1 & \boldsymbol{n}_2 & \boldsymbol{n}_3 \\ \dfrac{\partial}{\partial x_1} & \dfrac{\partial}{\partial x_2} & \dfrac{\partial}{\partial x_3} \\ A_1 & A_2 & A_3 \end{pmatrix} \tag{2.25}$$

with three components

$$\begin{aligned} D_1 &= \frac{\partial A_3}{\partial x_2} - \frac{\partial A_2}{\partial x_3} \\ D_2 &= \frac{\partial A_1}{\partial x_3} - \frac{\partial A_3}{\partial x_1} \\ D_3 &= \frac{\partial A_2}{\partial x_1} - \frac{\partial A_1}{\partial x_2}. \end{aligned} \tag{2.26}$$

Readers may verify using (2.10) and (2.11) that curl $\boldsymbol{A}$ transforms like an axial vector.

2.3 Tensors—definition

The word tensor literally means 'tension' because tensors were first introduced to describe the stress in a solid body. The stress tensor is now just a special case of a large class of tensors which occur in physics today.

A physical quantity, $\mathbf{T}$, with nine components (T_{ij}, $i, j = 1, 2, 3$) is called a tensor of rank two, if T_{ij} transform under an orthogonal transformation of axes like the products $(x_i x_j)$. Similarly, tensors of rank three (components T_{ijk}) and of rank four (components T_{ijkl}) have 27 and 81 components which transform, respectively, like the products $(x_i x_j x_k)$ and $(x_i x_j x_k x_l)$. Tensors of higher rank are defined in a similar way. From this point of view a vector is a tensor of rank one and a scalar is a tensor of rank zero.

Remembering the orthogonal transformation of the components in a Cartesian coordinate system (equations (2.1)), we have

$$x'_i x'_j = a_{im} x_m a_{jn} x_n = a_{im} a_{jn} x_m x_n. \tag{2.27}$$

Therefore, the transformation equations for the tensors of second rank are:

$$T'_{ij} = a_{im} a_{jn} T_{mn} \tag{2.28}$$

and, conversely,

$$T_{ij} = a_{mi} a_{nj} T'_{mn}. \tag{2.29}$$

Similarly,

$$T'_{ijk} = a_{im} a_{jn} a_{ko} T_{mno} \tag{2.30}$$

$$T_{ijk} = a_{mi} a_{nj} a_{ok} T'_{mno} \tag{2.31}$$

$$T'_{ijkl} = a_{im} a_{jn} a_{ko} a_{lp} T_{mnop} \tag{2.32}$$

$$T_{ijkl} = a_{mi} a_{nj} a_{ok} a_{pl} T'_{mnop} \tag{2.33}$$

and so on. In this book we will come across physical quantities which are tensors of up to rank four.

The usefulness of a physical quantity being identified as a tensor is twofold: (*a*) it immediately provides laws of transformation if we wish to work in other coordinate systems, and (*b*) if the problem has some inherent symmetry, the consequent simplification can be inferred easily as will become clear in due course. Tensors are useful in describing not only the elastic properties of materials, but many other properties as well. For example, the electrical conductivity, $\boldsymbol{\sigma}$, of a crystal is a symmetric tensor of rank two; it connects the electric field vector, $\boldsymbol{E}$, to the current density vector, $\boldsymbol{J}$, by

$$\boldsymbol{J} = \boldsymbol{\sigma} \boldsymbol{E} \qquad J_i = \sigma_{ij} E_j. \tag{2.34}$$

We shall see later that for a crystal of cubic symmetry one has

$$\sigma_{11} = \sigma_{22} = \sigma_{33} \qquad \text{and } \sigma_{ij} = 0 \text{ if } i \neq j \tag{2.35}$$

in all coordinate systems. Other examples of physical quantities which are described by symmetric tensors of rank two are the dielectric constant and the thermal expansion of a crystal (see §3.7).

2.4 Some elementary properties of tensors

In this section we enumerate some properties of tensors which will be useful later.

(i) If $\mathbf{T}_1$ and $\mathbf{T}_2$ are both tensors of rank m then their linear combination is also a tensor of rank m. Obviously,

$$\mathbf{T}=b_1\mathbf{T}_1+b_2\mathbf{T}_2 \tag{2.36}$$

where b_1 and b_2 are scalar constants, has the same transformation properties as $\mathbf{T}_1$ and $\mathbf{T}_2$.

(ii) Contraction rule: if $\mathbf{T}^{(m)}$ is a tensor of rank m with components $T_{ijkl\ldots m}$, then $T_{iikl\ldots m}$ are components of a tensor, $\mathbf{T}^{(m-2)}$, of rank $m-2$. To see this, note that

$$T'_{ijkl\ldots m}=a_{ir}a_{js}a_{kt}a_{lu}\ldots a_{mz}T_{rstu\ldots z} \tag{2.37}$$

and hence

$$\begin{aligned} T'_{iikl\ldots m}&\equiv\sum_{i=1}^{3}T'_{iikl\ldots m}\\ &=\left(\sum_{i=1}^{3}a_{ir}a_{is}\right)a_{kt}a_{lu}\ldots a_{mz}T_{rstu\ldots z}\\ &=a_{kt}a_{lu}\ldots a_{mz}T_{sstu\ldots z} \end{aligned} \tag{2.38}$$

since, from (2.2), the sum in brackets is just the Kronecker delta, δ_{rs}. Equation (2.38) shows that $T'_{iikl\ldots m}$ transform like the components of a tensor of rank $m-2$.

In particular, note that if $\mathbf{T}$ is a tensor of rank two then

$$T_{ii}=T_{11}+T_{22}+T_{33} \tag{2.39}$$

is a tensor of rank zero or scalar and hence invariant against the transformation of axes, i.e., $T'_{ii}=T_{ii}$.

The following properties refer to tensors of rank two.

(iii) Symmetric and antisymmetric tensors: if the elements, T_{ij}, of the tensor $\mathbf{T}$ are such that $T_{ij}=T_{ji}$, then it is called a symmetric tensor; and if $T_{ij}=-T_{ji}$ it is an antisymmetric tensor. For an antisymmetric tensor the diagonal elements T_{ii} are all zero.

A general tensor, $\mathbf{T}$, with elements $T_{ij}\neq T_{ji}$ can be written as the sum of the components of a symmetric and an antisymmetric tensor. Thus

$$\begin{aligned} T_{ij}&=\tfrac{1}{2}(T_{ij}+T_{ji})+\tfrac{1}{2}(T_{ij}-T_{ji})\\ &\equiv T_{ij}^{(s)}+T_{ij}^{(a)}. \end{aligned} \tag{2.40}$$

(iv) If $\boldsymbol{A}$ is a vector with components A_j, and $\mathbf{T}$ is a tensor with components T_{ij}, then

(*a*) $B_i\equiv T_{ij}A_j$ are components of a vector $\boldsymbol{B}$. To verify this consider

$$\begin{aligned} B'_i\equiv T'_{ij}A'_j&=\sum_{j=1}^{3}T'_{ij}A'_j=\sum_{j}\sum_{mn}(a_{im}a_{jn}T_{mn})\sum_{l}a_{jl}A_l\\ &=\sum_{j,m,n,l}\{a_{jn}a_{jl}\}a_{im}A_lT_{mn}. \end{aligned} \tag{2.41}$$

On using (2.2) equation (2.41) becomes

$$B_i' = \sum_{ml} a_{im} T_{ml} A_l = a_{im} B_m. \tag{2.42}$$

Thus $B_i \equiv T_{ij} A_j$ transform like the components of a vector.

(*b*) $C_i \equiv T_{ji} A_j$ are also components of a vector $\boldsymbol{C}$. Note that $\boldsymbol{B} = \boldsymbol{C}$ if $\mathbf{T}$ is a symmetric tensor.

(*c*) $\partial T_{ji}/\partial x_j$ are components of a vector: to see this consider

$$\frac{\partial T_{ji}'}{\partial x_j'} = \frac{\partial x_k}{\partial x_j'} \frac{\partial T_{ji}'}{\partial x_k}.$$

Using equations (2.1), (2.2) and (2.28) the above relation becomes

$$\frac{\partial T_{ji}'}{\partial x_j'} = a_{jk} \frac{\partial}{\partial x_k} (a_{jl} a_{im} T_{lm}) = a_{im} \frac{\partial T_{lm}}{\partial x_l} \tag{2.43}$$

showing that $\partial T_{ji}/\partial x_j$ transforms like the components of a vector. Similarly, $\partial T_{ij}/\partial x_j$ are also components of a vector.

(v) If $\boldsymbol{A}$ is a vector with components A_i, $i = 1, 2, 3$, then the nine quantities $\partial A_i/\partial x_j$ are components of a second-rank tensor, since

$$\frac{\partial A_i'}{\partial x_j'} = \frac{\partial x_k}{\partial x_j'} \frac{\partial}{\partial x_k} (a_{im} A_m) = a_{im} a_{jk} \frac{\partial A_m}{\partial x_k}. \tag{2.44}$$

It follows from (2.44) that

$$\frac{1}{2}\left(\frac{\partial A_i}{\partial x_j} + \frac{\partial A_j}{\partial x_i}\right) \equiv T_{ij}^{(s)} \tag{2.45}$$

and

$$\frac{1}{2}\left(\frac{\partial A_i}{\partial x_j} - \frac{\partial A_j}{\partial x_i}\right) \equiv T_{ij}^{(a)} \tag{2.46}$$

are, respectively, components of symmetric and antisymmetric tensors of rank two. We observe that $T_{ij}^{(a)} = \frac{1}{2}(\text{curl } \boldsymbol{A})_{ij}$, so that $T_{ij}^{(a)}$ also transform like the components of an axial vector. This identity of an antisymmetric tensor with an axial vector is true, however, only in three-dimensional space. For a general n-dimensional space, an antisymmetric tensor of rank two has $\nu = \frac{1}{2}n(n-1)$ components, while the vector has n components so that for

$$n = 1, 2, 3, 4, \ldots, \qquad \nu = 0, 1, 3, 6, \ldots,$$

and it is only for $n = 3$ that $\nu = n$.

2.5 Tensor quadric and principal axes of a symmetric tensor of rank two

As mentioned before symmetric tensors of rank two play an important role in physics. Such a tensor can be represented by a quadric surface as follows.

Let **S** be a symmetric tensor with components $S_{ij}\,(=S_{ji})$. If **S** represents an inherent physical property of a crystal then S_{ij} are naturally independent of the space coordinates. However, if **S** represents a strain or a stress, its values may vary from point to point in the solid; in this case it is important to remember that we are considering a quadric surface representation of the values of S_{ij} at a given point in the solid and hence in the following S_{ij} are just numbers (not functions of the coordinate variables occurring below).

Consider now a vector $\mathrm{OP} \equiv \boldsymbol{r}$ from the origin O of a Cartesian coordinate system to a point $\mathrm{P}(x_1, x_2, x_3)$. Remembering the results embodied in (2.42) and (2.17), we can associate with $\boldsymbol{r}$ and **S**, a vector $\boldsymbol{a}$

$$\boldsymbol{a} = \mathbf{S}\boldsymbol{r} \tag{2.47}$$

with components

$$a_i = S_{ij}x_j = S_{i1}x_1 + S_{i2}x_2 + S_{i3}x_3 \tag{2.48}$$

and a scalar $f(x_1, x_2, x_3)$:

$$\begin{aligned} f &= \boldsymbol{a}\cdot\boldsymbol{r} = a_i x_i = S_{ij}x_i x_j \\ &= S_{11}x_1^2 + S_{22}x_2^2 + S_{33}x_3^2 + 2S_{12}x_1x_2 + 2S_{13}x_1x_3 + 2S_{23}x_2x_3. \end{aligned} \tag{2.49}$$

The locus of points P such that $f(x_1, x_2, x_3) = \text{constant}$, is a quadric surface, ellipsoid or hyperboloid, depending on the values of S_{ij}. The constant is frequently set equal to unity and the surface

$$f = S_{ij}x_i x_j = 1 \tag{2.50}$$

is referred to, for brevity, as the tensor ellipsoid.

If we refer the quadric surface (2.50) to another coordinate system obtained by the orthogonal transformation (2.1), it would, in general, take the form

$$f = S'_{ij}x'_i x'_j = 1 \tag{2.51}$$

with S'_{ij} related to S_{ij} by the tensor transformation rule (2.28). There exists, however, at least one orthogonal coordinate system with axes parallel to X_1, X_2, X_3, say, referred to which

$$f = S_1X_1^2 + S_2X_2^2 + S_3X_3^2 = 1. \tag{2.52}$$

X_1, X_2, X_3 are the principal axes of the ellipsoid and S_1, S_2, S_3 are called the principal values of the tensor **S**. (The lengths of the principal axes are, of course, determined by the choice (here unity) of the constant in the equation $f = \text{constant}$.) We observe that referred to the principal axes, the tensor **S** is diagonal. The fact that now only three numbers (rather than the full six) seem to specify the tensor **S** is just because the other three numbers have gone into specifying the directions of the principal axes.

One way of determining the principal values and the directions of the principal axes is as follows.

First observe from equations (2.50) and (2.48) that

$$\tfrac{1}{2}\partial f/\partial x_i = S_{ij}x_j = a_i \tag{2.53}$$

or

$$\boldsymbol{a} = \tfrac{1}{2}\,\text{grad}\,f. \tag{2.54}$$

Hence the vector $\boldsymbol{a}$ defined in (2.47) in terms of the tensor **S** and vector OP $(=\boldsymbol{r})$ is normal to the ellipsoid at point P. If OP is taken along one of the principal axes—then it follows immediately from (2.52) and (2.54) that the corresponding $\boldsymbol{a}$ is parallel to OP or $\boldsymbol{r}$. In other words

$$\boldsymbol{a} = \lambda \boldsymbol{r} \qquad \boldsymbol{r}\text{ along a principal axis} \tag{2.55}$$

where λ is a constant. Using the coordinate system in which the components of $\boldsymbol{r}$ are x_1, x_2, x_3 and the components of $\boldsymbol{a}$ are given by (2.48), the vector equation (2.55) is equivalent to three linear equations in x_i:

$$S_{ij}x_j = \lambda x_i \qquad i = 1, 2, 3 \tag{2.56}$$

or, in full,

$$\begin{aligned}(S_{11}-\lambda)x_1 + S_{12}x_2 + S_{13}x_3 &= 0\\ S_{12}x_1 + (S_{22}-\lambda)x_2 + S_{23}x_3 &= 0\\ S_{13}x_1 + S_{23}x_2 + (S_{33}-\lambda)x_3 &= 0.\end{aligned} \tag{2.57}$$

These are simultaneous homogeneous equations and for a non-trivial solution the determinant formed from the coefficients of x_i must be zero:

$$\begin{vmatrix} S_{11}-\lambda & S_{12} & S_{13} \\ S_{12} & S_{22}-\lambda & S_{23} \\ S_{13} & S_{23} & S_{33}-\lambda \end{vmatrix} = 0. \tag{2.58}$$

This is a cubic equation in λ and its three roots λ_i, $i = 1, 2, 3$, are the three principal values S_i $(=\lambda_i)$ of the tensor **S**. The direction of a principal axis is determined by substituting the corresponding λ_i in equations (2.57) and solving them for the ratios $x_1 : x_2 : x_3$.

We note in passing that if the three principal values of **S** are equal, $S_1 = S_2 = S_3 = \lambda$, say, then referred to the principal axes, the quadric has the form

$$f = \lambda(X_1^2 + X_2^2 + X_3^2) \tag{2.59}$$

which transforms to

$$f = \lambda(x_1^2 + x_2^2 + x_3^2) \tag{2.60}$$

so that in all coordinate systems

$$S_{ik} = \lambda \delta_{ik} \tag{2.61}$$

for this case. Such a diagonal tensor with equal elements, when representing a physical quantity, signifies an isotropy. Consider, for example, the relation (2.34) between $\boldsymbol{J}$ and $\boldsymbol{E}$. For a general electrical conductivity tensor σ_{ij}, the electric current density would in general not be parallel to the applied electric field $\boldsymbol{E}$ unless $\boldsymbol{E}$ is along one of the principal axes of $\boldsymbol{\sigma}$. However, if $\boldsymbol{\sigma}$ has the form (2.61), namely $\sigma_{ij} = \lambda \delta_{ij}$, then $\boldsymbol{J} \| \boldsymbol{E}$ irrespective of the direction in which $\boldsymbol{E}$ is applied. We shall come across other examples later.

2.6 Invariants of a second-rank tensor

For a second-rank tensor $\mathbf{T}$, there exist certain combinations of its components, T_{ij}, which behave like scalars, that is, remain invariant under orthogonal transformations. We determine these invariants here for later use.

First, we show that the 3×3 determinant, det $\mathbf{T}$, with elements T_{ij} is a scalar, i.e.,

$$\det \mathbf{T} = \det \mathbf{T}' \tag{2.62}$$

where, from (2.28), $T'_{ij} = a_{im} a_{jn} T_{mn}$. Now both $\mathbf{T}$ and $\mathbf{a}$ are 3×3 square matrices. Hence, remembering the matrix multiplication rule for two matrices $\mathbf{A}$ and $\mathbf{B}$, namely,

$$(\mathbf{AB})_{mn} = \mathbf{A}_{mo} \mathbf{B}_{on} \tag{2.63}$$

we can write

$$T'_{ij} = a_{im} T_{mn} a_{jn} = (\mathbf{aT})_{in} (\mathbf{a}^{-1})_{nj} \tag{2.64}$$

or, in matrix form,

$$\mathbf{T}' = \mathbf{aTa}^{-1}. \tag{2.65}$$

Equation (2.65) implies

$$\det \mathbf{T}' = \det \mathbf{a} \det \mathbf{T} \det \mathbf{a}^{-1}$$

whence equation (2.62) follows since, from §2.1, det $\mathbf{a} = \det \mathbf{a}^{-1} = \pm 1$.

Now consider a second-rank tensor $\boldsymbol{\tau}$ whose elements are

$$\tau_{ij} = T_{ij} + \lambda \delta_{ij} \tag{2.66}$$

where λ is an arbitrary constant. The determinant of $\boldsymbol{\tau}$ is

$$\det \boldsymbol{\tau} = \begin{vmatrix} T_{11} + \lambda & T_{12} & T_{13} \\ T_{21} & T_{22} + \lambda & T_{23} \\ T_{31} & T_{32} & T_{33} + \lambda \end{vmatrix} \tag{2.67}$$

or

$$\det \boldsymbol{\tau} = \det \mathbf{T} + \lambda\Theta + \lambda^2\,\Delta + \lambda^3 \tag{2.68}$$

where

$$\Theta = T_{11}T_{22} + T_{22}T_{33} + T_{11}T_{33} - T_{12}T_{21} - T_{13}T_{31} - T_{23}T_{32} \tag{2.69}$$

and

$$\Delta = T_{ii} = T_{11} + T_{22} + T_{33}. \tag{2.70}$$

In a primed coordinate system the elements of the tensor are

$$\tau'_{ij} = T'_{ij} + \lambda\delta_{ij} \tag{2.71}$$

and hence

$$\det \boldsymbol{\tau}' = \det \mathbf{T}' + \lambda\Theta' + \lambda^2\,\Delta' + \lambda^3. \tag{2.72}$$

However, equation (2.62) is true for any second-rank tensor and hence

$$\det \boldsymbol{\tau}' = \det \boldsymbol{\tau} \tag{2.73}$$

also. Since (2.73) is true for any λ, equation (2.68) must equal (2.72) for all values of λ. Hence equating the coefficients of each power of λ, one has

$$\det \mathbf{T}' = \det \mathbf{T} \qquad \Delta' = \Delta \qquad \Theta = \Theta' \tag{2.74}$$

which are the required invariants. Of these, Δ, which we have come across earlier in equation (2.39), is sometimes called the first scalar or spur (sum of the diagonal elements) of a tensor. We note that any function of T_{ij} which is solely a function of the three invariants Δ, Θ and $\det \mathbf{T}$ is also a scalar.

If T is symmetric, the expressions for Θ and $\det \mathbf{T}$ given above are simplified by setting $T_{ij} = T_{ji}$. For an antisymmetric tensor, the diagonal elements T_{11} etc are zero and $T_{ij} = -T_{ji}$; hence $\Delta \equiv 0$, $\det \mathbf{T} \equiv 0$ and we have just one invariant, namely Θ, which is now

$$\Theta = T_{12}^2 + T_{13}^2 + T_{23}^2. \tag{2.75}$$

Problems

2.1 Let a primed coordinate system (x'_1, x'_2, x'_3) be obtained from the unprimed coordinate system by rotation of the angle φ about the x_3-axis. Verify that the transformation matrix a_{ij} is

$$a_{ij} = \begin{pmatrix} \cos\varphi & \sin\varphi & 0 \\ -\sin\varphi & \cos\varphi & 0 \\ 0 & 0 & 1 \end{pmatrix} \tag{P2.1}$$

2.2 A primed coordinate system (x'_1, x'_2, x'_3) is obtained from the unprimed system by the following successive rotations: (i) rotate unprimed axes through angle α about the x_3-axis to obtain the axes $x''_1, x''_2, x''_3\ (\equiv x_3)$; (ii) rotate the x''_i-axes about the x''_2-axis through an angle β to obtain $x'''_1, x'''_2\ (\equiv x''_2)$ and x'''_3 axes; (iii) rotate the x'''_i-axes through angle γ about the x'''_3-axis to obtain x'_1-, x'_2-, x'_3-axes. Verify that the transformation matrix a_{ij} for $x'_i = a_{ij}x_j$ is

$$\begin{pmatrix} \cos\alpha\cos\beta\cos\gamma - \sin\alpha\sin\gamma & \sin\alpha\cos\beta\cos\gamma + \cos\alpha\sin\gamma & -\sin\beta\cos\gamma \\ -\cos\alpha\cos\beta\sin\gamma - \sin\alpha\cos\gamma & -\sin\alpha\cos\beta\sin\gamma + \cos\alpha\cos\gamma & \sin\beta\sin\gamma \\ \cos\alpha\sin\beta & \sin\alpha\sin\beta & \cos\beta \end{pmatrix} \tag{P2.2}$$

All rotations are in the right-hand screw sense. α, β, γ are known as the Euler angles. (See, for example, Mathews and Walker (1970) p 404.) (P2.2) represents the most general transformation for rotations† and will be useful later.

2.3 (*a*) Let $\boldsymbol{A}$ be an ordinary (polar) vector. Verify using the transformation matrices (2.10) and (2.11) that curl $\boldsymbol{A}$ is an axial vector and curl curl $\boldsymbol{A}$ is a polar vector.

(*b*) If $\boldsymbol{A}$ is an axial vector how do curl $\boldsymbol{A}$ and curl curl $\boldsymbol{A}$ transform?

2.4 Let $T_{ijkl\ldots n}$ be a tensor of rank n in three-dimensional space $(i, j, k, \ldots, n = 1, 2, 3)$. Show that the following are tensors of rank $n-2$:

$$X_{k\ldots n} \equiv T_{iikl\ldots n}$$

$$Y_{k\ldots n} \equiv (T_{ijkl\ldots n})(A_i B_j)$$

$$Z_{k\ldots n} \equiv (T_{ijkl\ldots n})(\tau_{ij})$$

where A_i and B_j are components of the vectors $\boldsymbol{A}$ and $\boldsymbol{B}$ and τ_{ij} are components of a tensor of rank two.

2.5 Consider the cylindrical polar coordinates (r, φ and $z\ (\equiv x_3)$) at a point P. $r = (x_1^2 + x_2^2)^{1/2}$, $x_1 = r\cos\varphi$, $x_2 = r\sin\varphi$. Set up a primed coordinate system with $x'_1 \parallel \boldsymbol{r}$, $x'_2 \parallel r\,\mathrm{d}\varphi$ direction and $x'_3 = x_3$.

(*a*) Find the components, T'_{ij}, of a symmetric second-rank tensor $\mathbf{T}$ in terms of the components T_{ij} and *vice versa*.

(*b*) Do the same as (*a*) if $\mathbf{T}$ is antisymmetric. (Note: T'_{ij} of this problem are often written as T_{rr}, $T_{r\varphi}$, $T_{\varphi z}$, T_{zz} etc.)

2.6 Consider spherical polar coordinates (r, θ, φ) at a point P:

$$x_1 = r\sin\theta\cos\varphi \qquad x_2 = r\sin\theta\sin\varphi \qquad x_3 = r\cos\theta.$$

Set up a primed coordinate system at P so that

$$x'_3 \parallel \boldsymbol{r} \qquad x'_1 \parallel r\,\mathrm{d}\theta \text{ direction}$$

† The choice of successive rotations sometimes varies in the literature and therefore the matrix (P2.2) may look different. For discussion see Goldstein (1980), p. 143.

and

$$x'_2 \| r \sin\theta \, \mathrm{d}\varphi \text{ direction.}$$

(*a*) Determine the matrix a_{ij} using (P2.2) or otherwise. (Note: set $\gamma=0$, $\beta\equiv\theta$, $\alpha\equiv\varphi$.)
(*b*) Find the components T'_{ij} in terms of T_{ij} and *vice versa*.

2.7 A symmetric tensor S_{ij} has the following components in a given coordinate system (x_1, x_2, x_3):

$$\mathbf{S}=\begin{pmatrix} 5 & 7 & 0 \\ 7 & 3 & 0 \\ 0 & 0 & 4 \end{pmatrix}.$$

(*a*) Determine the principal values and the directions of the principal axes of the tensor.
(*b*) Determine the three invariants of the tensor.

2.8 Do the same problem as **2.7** when

$$S_{ij}=\begin{pmatrix} 8 & 0 & -\sqrt{2} \\ 0 & 6 & 0 \\ -\sqrt{2} & 0 & 7 \end{pmatrix}.$$

(Comment: In this case two of the principal values are equal; hence the corresponding two principal axes can be taken along any two mutually orthogonal lines lying in the plane perpendicular to the third principal axis.)

2.9 Determine all the invariants of a tensor of rank two in two-dimensional space. What do these expressions reduce to if the tensor is (*a*) symmetric and (*b*) antisymmetric?

2.10 Show that
(i) in cylindrical polar coordinates (r, φ, z) that:

$$\nabla\cdot\boldsymbol{A}=\frac{1}{r}\frac{\partial}{\partial r}rA_r+\frac{1}{r}\frac{\partial A_\varphi}{\partial\varphi}+\frac{\partial A_z}{\partial z} \tag{P2.3}$$

$$(\nabla\times\boldsymbol{A})_r=\frac{1}{r}\frac{\partial A_z}{\partial\varphi}-\frac{\partial A_\varphi}{\partial z} \tag{P2.4}$$

$$(\nabla\times\boldsymbol{A})_\varphi=\frac{\partial A_r}{\partial z}-\frac{\partial A_z}{\partial r} \tag{P2.5}$$

$$(\nabla\times\boldsymbol{A})_z=\frac{1}{r}\frac{\partial}{\partial r}rA_\varphi-\frac{1}{r}\frac{\partial A_r}{\partial\varphi}. \tag{P2.6}$$

(ii) in spherical polar coordinates (r, θ, φ) that

$$\nabla \cdot \boldsymbol{A} = \frac{1}{r^2}\frac{\partial}{\partial r} r^2 A_r + \frac{1}{r\sin\theta}\frac{\partial}{\partial\theta}(\sin\theta A_\theta) + \frac{1}{r\sin\theta}\frac{\partial A_\varphi}{\partial\varphi} \tag{P2.7}$$

$$(\nabla \times \boldsymbol{A})_r = \frac{1}{r\sin\theta}\left(\frac{\partial}{\partial\theta}(\sin\theta A_\varphi) - \frac{\partial A_\theta}{\partial\varphi}\right) \tag{P2.8}$$

$$(\nabla \times \boldsymbol{A})_\theta = \frac{1}{r\sin\theta}\frac{\partial A_r}{\partial\varphi} - \frac{1}{r}\frac{\partial}{\partial r} r A_\varphi \tag{P2.9}$$

$$(\nabla \times \boldsymbol{A})_\varphi = \frac{1}{r}\left(\frac{\partial}{\partial r} r A_\theta - \frac{\partial A_r}{\partial\theta}\right). \tag{P2.10}$$

3

DEFORMATION, STRAIN AND THERMAL EXPANSION TENSORS

When a body is stretched, twisted or deformed in any other way each element of matter in it is, in general, displaced from its original position in the undeformed state. This Chapter is mainly concerned with the problem of specifying these deformations. To begin with it is useful to consider a simple example of deformation in one dimension.

3.1 Deformation in one dimension

Consider a straight thin piece of wire in which, as an idealisation, we neglect all dimensions transverse to its length. Then the deformation consists of a simple extension or contraction. We denote the position of a particle in the wire in its undeformed state by x and the position of the same particle in a deformed state by l ($\equiv l(x)$). If l is known for all x, the deformation is completely specified. The displacement of the particle originally at x is

$$s(x)=l(x)-x. \tag{3.1}$$

If $s(x)$ is constant, independent of x, then particles at all points of the wire are displaced by the same amount; we then have no deformation but only a simple translation of the wire.

Now consider two particles of the wire with coordinates x and $x+\mathrm{d}x$ and located at points P and Q, as in figure 3.1. On deformation the material at P moves to P′ and the material at Q to Q′ with coordinates

$$l(x)=x+s(x) \qquad l(x+\mathrm{d}x)=x+\mathrm{d}x+s(x+\mathrm{d}x). \tag{3.2}$$

The length of the infinitesimal element PQ after deformation is

$$\mathrm{P'Q'}=\mathrm{d}l=l(x+\mathrm{d}x)-l(x)=\mathrm{d}x+s(x+\mathrm{d}x)-s(x). \tag{3.3}$$

On making a Taylor expansion of $s(x+\mathrm{d}x)$ and retaining only the lowest power of the infinitesimal $\mathrm{d}x$, one has

$$\mathrm{d}l=\mathrm{d}x+(\mathrm{d}s/\mathrm{d}x)\,\mathrm{d}x\equiv\mathrm{d}x(1+\varepsilon) \tag{3.4}$$

where

$$\varepsilon=\frac{\mathrm{d}s}{\mathrm{d}x}=\frac{\mathrm{d}l-\mathrm{d}x}{\mathrm{d}x} \tag{3.5}$$

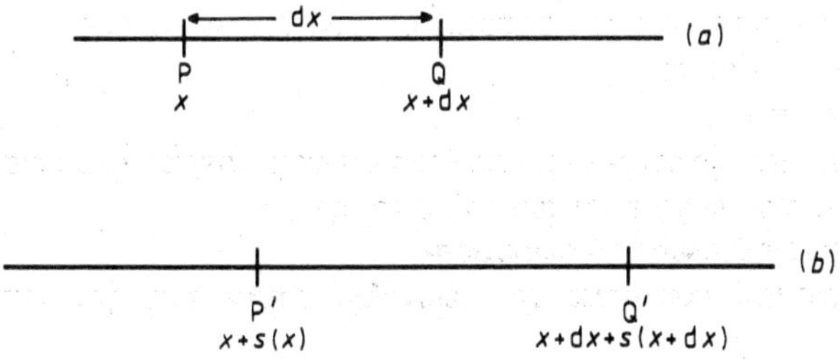

Figure 3.1 The position of the two elements in the wire: (*a*) in the undeformed state and (*b*) after deformation.

is the fractional change in the length of an element of the wire at x and is usually known as the strain in the wire at point x.

If ε is constant throughout the length of the wire, the strain is called *homogeneous*. For a homogeneous strain, integrating equation (3.4), one has

$$L=(1+\varepsilon)L_0 \qquad \text{or } \varepsilon=(L-L_0)/L_0 \tag{3.6}$$

where L_0 and L are the lengths of the wire before and after the deformation.

3.2 Deformation in three dimensions

3.2.1 Geometry of the deformation and deformation tensor

The specification of the deformation of a region around a given point in a three-dimensional body is naturally more involved since both the positions of the various points (particles) in the region and their displacements are vectors. Consider again the infinitesimal line element $\mathrm{d}\boldsymbol{x}$, depicted as PQ in figure 3.2, before deformation. The position vectors of P and Q with reference to a fixed set of axes in space are $\boldsymbol{x}$ and $\boldsymbol{x}+\mathrm{d}\boldsymbol{x}$. After deformation due to a displacement field $\boldsymbol{s}(\boldsymbol{x})$, the material element PQ occupies the position P′Q′, the coordinates of P′ and Q′ being $\boldsymbol{l}(\boldsymbol{x})$ and $\boldsymbol{l}(\boldsymbol{x}+\mathrm{d}\boldsymbol{x})$ respectively with

$$\boldsymbol{l}(\boldsymbol{x})=\boldsymbol{x}+\boldsymbol{s}(\boldsymbol{x}) \qquad \boldsymbol{l}(\boldsymbol{x}+\mathrm{d}\boldsymbol{x})=\boldsymbol{x}+\mathrm{d}\boldsymbol{x}+\boldsymbol{s}(\boldsymbol{x}+\mathrm{d}\boldsymbol{x}). \tag{3.7}$$

From the geometry (see figure 3.2), the vector is given by

$$\mathrm{P'Q'}\equiv\mathrm{d}\boldsymbol{l}=\boldsymbol{l}(\boldsymbol{x}+\mathrm{d}\boldsymbol{x})-\boldsymbol{l}(\boldsymbol{x})=\mathrm{d}\boldsymbol{x}+\boldsymbol{s}(\boldsymbol{x}+\mathrm{d}\boldsymbol{x})-\boldsymbol{s}(\boldsymbol{x}). \tag{3.8}$$

On making a Taylor expansion of $\boldsymbol{s}(\boldsymbol{x}+\mathrm{d}\boldsymbol{x})$ around $\boldsymbol{x}$ and confining our attention to a small enough region so that squares and products of $\mathrm{d}x_i$ can be neglected, we have for the components $\mathrm{d}l_i$ $(i=1, 2, 3)$ of $\mathrm{d}\boldsymbol{l}$:

$$\mathrm{d}l_1=\mathrm{d}x_1+\frac{\partial s_1}{\partial x_1}\,\mathrm{d}x_1+\frac{\partial s_1}{\partial x_2}\,\mathrm{d}x_2+\frac{\partial s_1}{\partial x_3}\,\mathrm{d}x_3 \tag{3.9a}$$

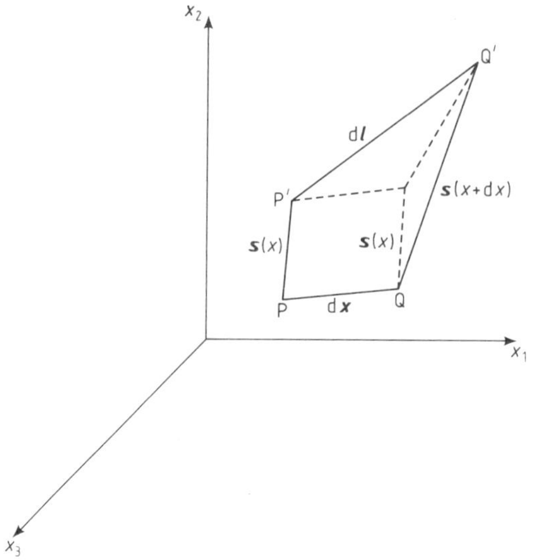

Figure 3.2 Figure showing the position of an element before deformation (PQ) and after deformation (P′Q′).

$$dl_2 = dx_2 + \frac{\partial s_2}{\partial x_1}\,dx_1 + \frac{\partial s_2}{\partial x_2}\,dx_2 + \frac{\partial s_2}{\partial x_3}\,dx_3 \qquad (3.9b)$$

$$dl_3 = dx_3 + \frac{\partial s_3}{\partial x_1}\,dx_1 + \frac{\partial s_3}{\partial x_2}\,dx_2 + \frac{\partial s_3}{\partial x_3}\,dx_3 \qquad (3.9c)$$

or, in abbreviated form:

$$dl_i = \left(\delta_{ij} + \frac{\partial s_i}{\partial x_j}\right) dx_j. \qquad (3.10)$$

The above shows that the effect of the displacement $\boldsymbol{s}$ on an infinitesimal region† around P can be considered as a sum of two parts: a translation of the entire region by an amount $\boldsymbol{s}(\boldsymbol{x})$ which obviously constitutes no deformation of the region, and a change in length and orientation of every line element PQ originating from P, the element occupying the position P′Q′ after the deformation. The components dl_i of the vector P′Q′ depend linearly on the components dx_i of PQ and on $\partial s_i/\partial x_j$ evaluated at $\boldsymbol{x}$. The nine quantities $\partial s_i/\partial x_j$ thus determine the deformation of the region and according to equation (2.44) are components of a second-rank tensor. This tensor, with components $\partial s_i/\partial x_j$, is called the deformation tensor at $\boldsymbol{x}$.

† We note that the region is, nonetheless, assumed to contain many atoms so that a translation or rotation (see below) of the region as a whole is effectively like that of a rigid body where no interatomic forces come in to play in equilibrium considerations.

When all the $\partial s_i/\partial x_j$ are constants, independent of $\boldsymbol{x}$, throughout the body, the deformation is called homogeneous. For a homogeneous deformation, equations (3.9) and (3.10) can be integrated and they apply equally to the line elements of any length. Thus if L_0 is the length of an element before deformation and n_1, n_2, n_3 represents its direction cosines then the components, L_i, of the element in the deformed state of the body may be obtained from (3.10) as (all $\partial s_i/\partial x_j$ are constants)

$$L_i = L_0[\delta_{ij} + (\partial s_i/\partial x_j)]n_j. \tag{3.11}$$

The deformation tensor includes a rotation of the small region as a rigid body—or, for a homogeneous deformation rotation of the whole specimen as a rigid body. For the important case of infinitesimal deformations, that is, where $\partial s_i/\partial x_j \ll 1$, the delineation of this rotation is quite simple as we show below.

At this point we mention that in this book we shall be concerned only with the theory of elasticity of solids for infinitesimal deformations. The subject of the elasticity of solids for finite deformations, besides being more complex, is today almost as extensive and has applications, among others, in polymer physics and rubber elasticity; the reader interested in this is referred to such treatises as Murnaghan (1951), Green and Zerna (1954) and Green and Adkins (1970). The question of specifying the strain for finite deformations is, however, in itself instructive and is briefly discussed in §3.4.

3.2.2 The infinitesimal strain and rotation tensors

Let us decompose the deformation tensor into its symmetric (ε_{ij}) and antisymmetric (t_{ij}) parts:

$$\varepsilon_{ij} = \varepsilon_{ji} = \frac{1}{2}\left(\frac{\partial s_i}{\partial x_j} + \frac{\partial s_j}{\partial x_i}\right) \tag{3.12}$$

and

$$t_{ij} = -t_{ji} = -\frac{1}{2}\left(\frac{\partial s_i}{\partial x_j} - \frac{\partial s_j}{\partial x_i}\right). \tag{3.13}$$

The t_{ij} may be regarded as the components of an axial vector $\boldsymbol{R} \equiv \frac{1}{2}$ curl $\boldsymbol{s}$, so that

$$R_1 = t_{23} = \frac{1}{2}\left(\frac{\partial s_3}{\partial x_2} - \frac{\partial s_2}{\partial x_3}\right) \tag{3.14a}$$

$$R_2 = t_{31} = \frac{1}{2}\left(\frac{\partial s_1}{\partial x_3} - \frac{\partial s_3}{\partial x_1}\right) \tag{3.14b}$$

$$R_3 = t_{12} = \frac{1}{2}\left(\frac{\partial s_2}{\partial x_1} - \frac{\partial s_1}{\partial x_2}\right). \tag{3.14c}$$

Using (3.12)–(3.14), we can rewrite equation (3.9*a*) for $\mathrm{d}l_1$ as

$$\mathrm{d}l_1 = \mathrm{d}x_1 + \frac{\partial s_1}{\partial x_1}\,\mathrm{d}x_1 + \frac{1}{2}\left(\frac{\partial s_1}{\partial x_2} + \frac{\partial s_2}{\partial x_1}\right)\mathrm{d}x_2 + \frac{1}{2}\left(\frac{\partial s_1}{\partial x_3} + \frac{\partial s_3}{\partial x_1}\right)\mathrm{d}x_3$$

$$+\frac{1}{2}\left(\frac{\partial s_1}{\partial x_2} - \frac{\partial s_2}{\partial x_1}\right)\mathrm{d}x_2 + \frac{1}{2}\left(\frac{\partial s_1}{\partial x_3} - \frac{\partial s_3}{\partial x_1}\right)\mathrm{d}x_3$$

$$= \mathrm{d}x_1 + \varepsilon_{11}\,\mathrm{d}x_1 + \varepsilon_{12}\,\mathrm{d}x_2 + \varepsilon_{13}\,\mathrm{d}x_3 - R_3\,\mathrm{d}x_2 + R_2\,\mathrm{d}x_3. \qquad (3.15)$$

Similarly rewriting $\mathrm{d}l_2$ and $\mathrm{d}l_3$, one sees that $\mathrm{d}\boldsymbol{l}$ can be written in vector form as the sum of three vectors:

$$\mathrm{d}\boldsymbol{l} = \mathrm{d}\boldsymbol{x} + \mathrm{d}\boldsymbol{l}' + \mathrm{d}\boldsymbol{l}'' \qquad (3.16)$$

where

$$\mathrm{d}\boldsymbol{l}' = \boldsymbol{\varepsilon}\,\mathrm{d}\boldsymbol{x} \qquad \mathrm{d}l_i' = \varepsilon_{ij}\,\mathrm{d}x_j \qquad (3.17)$$

and

$$\mathrm{d}\boldsymbol{l}'' = \boldsymbol{R} \times \mathrm{d}\boldsymbol{x}. \qquad (3.18)$$

The square of the length of the element in the deformed state is

$$(\mathrm{P'Q'})^2 = |\mathrm{d}l|^2 = \mathrm{d}\boldsymbol{l}\cdot\mathrm{d}\boldsymbol{l} = \mathrm{d}\boldsymbol{l}\cdot(\mathrm{d}\boldsymbol{x} + \mathrm{d}\boldsymbol{l}' + \mathrm{d}\boldsymbol{l}'')$$

$$\simeq |\mathrm{d}x|^2 + 2\,\mathrm{d}\boldsymbol{x}\cdot\mathrm{d}\boldsymbol{l}' \qquad (3.19)$$

$$= |\mathrm{d}x|^2 + 2\varepsilon_{ij}\,\mathrm{d}x_i\,\mathrm{d}x_j \qquad (3.20)$$

where we have neglected squares and products of ε_{ij} and R_i on the grounds that for infinitesimal deformations all $\partial s_i/\partial x_j \ll 1$, and so only linear terms in them need be retained. Note that the scalar product $\mathrm{d}\boldsymbol{x}\cdot\mathrm{d}\boldsymbol{l}''$ is identically zero.

Equation (3.20) shows that if all $\varepsilon_{ij} \equiv 0$, then $|\mathrm{d}l| = |\mathrm{d}x|$ for all lengths and orientations of the infinitesimal element $\mathrm{d}\boldsymbol{x}$. This means that the deformation tensor for this case, that is, t_{ij} or $\boldsymbol{R}$, represents a pure rotation of an infinitesimal region† around $\boldsymbol{x}$ as a rigid body. The symmetric tensor ε_{ij} which alone is responsible for changes in shape or volume of the region is known as the strain tensor 'for infinitesimal deformations'—the words within inverted commas will be omitted for brevity in future. We consider the properties of the strain tensor in a separate section, concluding this section by making a few remarks on the rotation tensor which will be useful later.

3.2.3 *The rotation vector* $\boldsymbol{R}$ *and the vortex vector*

As an example of how $\boldsymbol{R}$ changes the orientation of a line element, suppose that $\boldsymbol{R}$ is along the x_3-axis of our coordinate system and the line element PQ_1 is

† For a homogeneous deformation, of course, t_{ij} or $\boldsymbol{R}$ represents a rigid rotation of the whole specimen.

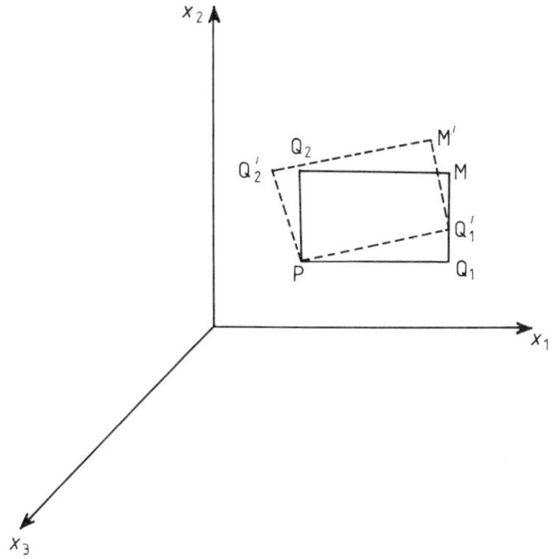

Figure 3.3 Displacement of the rectangle PQ_1MQ_2P in the x_1–x_2 plane due to the rotation $\boldsymbol{R}$ with components $R_1=R_2=0$ and $R_3\neq 0$ (the rectangle is rotated as a rigid area in the anticlockwise direction through an angle $\varphi \simeq R_3$).

along the x_1-axis (figure 3.3). Since $R_1=R_2=0$, $R_3\neq 0$, and $dx_2=dx_3=0$, $dx_1\neq 0$, one has from (3.18)

$$(dl'')_1=(dl'')_3=0 \qquad (dl'')_2=R_3\, dx_1.$$

Thus the element PQ_1 goes to PQ_1' which is obtained by an anticlockwise rotation through an angle φ about the x_3-axis,

$$\text{tam } \varphi \simeq \varphi = \frac{R_3\, dx_1}{dx_1} = R_3 \tag{3.21}$$

since R_3 and hence tan φ and φ are small compared with unity. To linear terms in R_3 the length of the element $PQ_1'=[(dx_1)^2+(R_3\, dx_1)^2]^{1/2}\simeq dx_1$ as it should be. Similarly for an element PQ_2 $(0, dx_2, 0)$ along the x_2-axis:

$$(dl'')_1=-R_3\, dx_2 \qquad (dl'')_2=(dl'')_3=0$$

so that the effect of R_3 on PQ_2 is again a rotation in the anticlockwise direction through an angle φ about the x_3-axis: $\tan \varphi \simeq \varphi = R_3\, dx_2/dx_2 = R_3$.

Quite generally, one may verify that the displacement (3.18) represents an infinitesimal rotation of the element $d\boldsymbol{x}$ by angle $|R|$ about an axis parallel to $\boldsymbol{R}$ and passing through P. As already mentioned the tensor t_{ij} or the vector $\boldsymbol{R}$ represents a rotation when $\partial s_i/\partial x_j \ll 1$. The case of rotations when $\partial s_i/\partial x_j$ are not small will not be discussed here (see, for example, Love 1944).

Though infinitesimal rotations do not play a role in static problems, they are, nonetheless, of importance when we consider the time dependence of the displacements. If $\boldsymbol{v}$ is the velocity of an element at $\boldsymbol{x}$, a small displacement $\boldsymbol{s}=\boldsymbol{v}\,\Delta t$ occurs in the small time Δt. Then since $s_i=v_i\Delta t$ we have

$$\boldsymbol{R}=\tfrac{1}{2}\,\mathrm{curl}\,\boldsymbol{s}=\tfrac{1}{2}\,\mathrm{curl}(\boldsymbol{v}\,\Delta t)=\Delta t(\tfrac{1}{2}\,\mathrm{curl}\,\boldsymbol{v}).$$

Hence setting

$$\boldsymbol{w}=\tfrac{1}{2}\,\mathrm{curl}\,\boldsymbol{v} \qquad (3.22)$$

one obtains for the rotation in a time Δt, $\boldsymbol{R}=\boldsymbol{w}\,\Delta t$. $\boldsymbol{w}$ is called the vortex vector or vorticity. We shall see later that it plays an important role in the description of the motion of fluids as was first discussed by Helmholtz (see §10.4).

3.3 The strain tensor

3.3.1 Strain quadric

Having seen that the antisymmetric part (t_{ij} or $\boldsymbol{R}$) of the deformation tensor at point P describes an infinitesimal rotation of a small region around P as a rigid body, we now return to the symmetric part, that is, the strain tensor ε_{ij}. The components of the displacement $\mathrm{d}\boldsymbol{l}'$ due to ε_{ij} are given by (3.17), namely, $\mathrm{d}l'_i=\varepsilon_{ij}\,\mathrm{d}x_j$. Some insight into the strain tensor is gained immediately if we form the corresponding tensor quadric f and refer it to its principal axis in §2.5†:

$$f\equiv\mathrm{d}\boldsymbol{l}'\cdot\mathrm{d}\boldsymbol{x}=\mathrm{d}l'_i\,\mathrm{d}x_i=\varepsilon_{ij}\,\mathrm{d}x_i\,\mathrm{d}x_j=\mathrm{const.} \qquad (3.23)$$

We have

$$\mathrm{d}l'_i=\tfrac{1}{2}\,\partial f/\partial(\mathrm{d}x_i)=\varepsilon_{ij}\,\mathrm{d}x_j \qquad (3.24)$$

as in (3.17).

Denoting the principal axes of the quadric by $\mathrm{d}X_1$, $\mathrm{d}X_2$ and $\mathrm{d}X_3$ and the principal values of the strain tensor (at x) by $\varepsilon^{(1)}$, $\varepsilon^{(2)}$, $\varepsilon^{(3)}$, we have

$$f=\varepsilon_{ij}\,\mathrm{d}x_i\,\mathrm{d}x_j=\varepsilon^{(1)}(\mathrm{d}X_1)^2+\varepsilon^{(2)}(\mathrm{d}X_2)^2+\varepsilon^{(3)}(\mathrm{d}X_3)^2. \qquad (3.25)$$

Hence the components of $\mathrm{d}\boldsymbol{l}'$ along the principal axes are:

along X_1: $$(\mathrm{d}l')_{X_1}=\tfrac{1}{2}[\partial f/\partial(\mathrm{d}X_1)]=\varepsilon^{(1)}\,\mathrm{d}X_1 \qquad (3.26a)$$

along X_2: $$(\mathrm{d}l')_{X_2}=\tfrac{1}{2}[\partial f/\partial(\mathrm{d}X_2)]=\varepsilon^{(2)}\,\mathrm{d}X_2 \qquad (3.26b)$$

† In §2.5 we worked with x_j, rather than $\mathrm{d}x_j$, but obviously we can equally well work with $\mathrm{d}x_j$ since $\mathrm{d}\boldsymbol{x}$ is also a spatial vector.

along X_3: $$(\mathrm{d}l')_{X_3} = \tfrac{1}{2}[\partial f/\partial(\mathrm{d}X_3)] = \varepsilon^{(3)}\,\mathrm{d}X_3. \tag{3.26c}$$

Thus, if we take our line element PQ along one of the principal axes of the strain tensor at P, say along the X_1-axis and of length $\mathrm{d}X_1$, then due to the strain its length is changed to

$$\mathrm{d}X_1 + (\mathrm{d}l')_{X_1} = \mathrm{d}X_1(1+\varepsilon^{(1)}) \tag{3.27}$$

which constitutes a simple extension or contraction, depending on the sign of $\varepsilon^{(1)}$, along its length. Similarly, the lengths of the line elements $\mathrm{d}X_2$ and $\mathrm{d}X_3$ along the other two principal axes become $\mathrm{d}X_2(1+\varepsilon^{(2)})$ and $\mathrm{d}X_3(1+\varepsilon^{(3)})$ respectively. Thus, the deformation of a small region around P due to strain may be described as a simple extension or contraction along three mutually perpendicular directions parallel to the principal axes of the strain tensor at P.

We consider some other properties of the strain tensor in §§3.3.2 and 3.3.3.

3.3.2 Fractional volume change of a small parallelepiped (dilatation)

Consider a rectangular parallelepiped of sides $\mathrm{d}X_1$, $\mathrm{d}X_2$, $\mathrm{d}X_3$ parallel to the three principal axes of the strain tensor ε_{ij}. In the strained state the lengths of the sides become $\mathrm{d}X_i(1+\varepsilon^{(i)})$, $i=1, 2, 3$, and hence the change in the volume of the parallelepiped is

$$\Delta V' - \Delta V = \mathrm{d}X_1\,\mathrm{d}X_2\,\mathrm{d}X_3(1+\varepsilon^{(1)})(1+\varepsilon^{(2)})(1+\varepsilon^{(3)}) - \mathrm{d}X_1\,\mathrm{d}X_2\,\mathrm{d}X_3.$$

Hence the fractional change in volume, called the dilatation, is

$$\begin{aligned}\Delta \equiv (\Delta V' - \Delta V)/\Delta V &= (1+\varepsilon^{(1)})(1+\varepsilon^{(2)})(1+\varepsilon^{(3)}) - 1\\ &= \varepsilon^{(1)} + \varepsilon^{(2)} + \varepsilon^{(3)}\end{aligned} \tag{3.28}$$

to linear terms in $\varepsilon^{(i)}$. We know from §2.6 that the diagonal sum of a second-rank tensor is invariant against coordinate transformations. Hence, in any orthogonal coordinate system, the dilatation is

$$\Delta = \varepsilon_{ii} = \varepsilon_{11} + \varepsilon_{22} + \varepsilon_{33}. \tag{3.29}$$

Using (3.12) in (3.29), we also have

$$\Delta = \frac{\partial s_i}{\partial x_i} = \frac{\partial s_1}{\partial x_1} + \frac{\partial s_2}{\partial x_2} + \frac{\partial s_3}{\partial x_3} \equiv \operatorname{div} \boldsymbol{s} \tag{3.30}$$

a relation which will prove useful later.

For the case of uniform compression or expansion, such as that occurring in a fluid under hydrostatic pressure, the strain tensor has the form

$$\varepsilon_{ij} = \tfrac{1}{3}\,\Delta\delta_{ij}. \tag{3.31}$$

For this case the strain quadric is a sphere (cf equations (2.59)–(2.61)) so that the extension (or contraction) of every line element, irrespective of its direction, is the same, namely $\tfrac{1}{3}\,\Delta$.

3.3.3 Interpretation of the individual components ($\varepsilon_{11}, \varepsilon_{12}$ etc) of the strain tensor

(*a*) Longitudinal strain ($\varepsilon_{ij}, i=j$)
Consider a long fibre, lying along the x_1-axis such as a piece of wire of length dx_1 (long in the sense that its transverse dimensions are small compared with its length dx_1) in the Young's modulus experiment. Then the vector $d\boldsymbol{x}$ of the fibre has components dx_1, 0, 0, and the square of its length in the strained state from equation (3.20) is

$$|d\boldsymbol{l}|^2=(dx_1)^2+2\varepsilon_{11}(dx_1)^2=(dx_1)^2(1+2\varepsilon_{11}). \tag{3.32}$$

Hence remembering that we are throughout considering infinitesimal strains ($\varepsilon_{11} \ll 1$), we have

$$(|d\boldsymbol{l}|-dx_1)/dx_1=(1+2\varepsilon_{11})^{1/2}-1 \simeq \varepsilon_{11}. \tag{3.33}$$

Thus ε_{11} is simply the extension, per unit length, of a thin fibre lying along the x_1-axis. A similar interpretation applies to ε_{22} and ε_{33}.

It hardly needs be added that for homogeneous strain the formula (3.33) applies not only to wires of infinitesimal length dx_1 but also to those of finite lengths.

The above interpretation of ε_{11}, in conjunction with the transformation properties of the strain tensor, may be used to determine the extension of the thin fibre which lies along an arbitrary direction in space, say, having direction cosines n_1, n_2, n_3. For, if we set up a primed Cartesian coordinate system whose x_1'-axis is parallel to the fibre, the extension, per unit length, of the fibre is just ε_{11}', where ε_{ij}' are the components of the strain tensor in the primed system. From equation (2.28) we can write

$$\varepsilon_{ij}'=a_{ik}a_{jl}\varepsilon_{kl} \tag{3.34}$$

and hence

$$\varepsilon_{11}'=a_{1k}a_{1l}\varepsilon_{kl}=n_k n_l \varepsilon_{kl} \tag{3.35}$$

since an element a_{1k}, $k=1, 2, 3$, of the transformation matrix equals n_k. The same extension for the fibre may, of course, be obtained from (3.20) by remembering that if dr is the length of the fibre, $dx_i \equiv n_i\, dr$ and following the argument leading up to (3.33).

(*b*) Shear strain ($\varepsilon_{ij}, i\neq j$)
To interpret ε_{12}, let all other components of the strain tensor be zero and consider the effect of ε_{12} on two line elements PQ_1 and PQ_2 parallel, respectively, to the x_1- and x_2-axes and of lengths dx_1 and dx_2 (figure 3.4). In the strained state due to ε_{12}, PQ_1 occupies the position PQ_1' and PQ_2 the position PQ_2'. Denoting these vectors by $d\boldsymbol{l}$ and $d\bar{\boldsymbol{l}}$ respectively, their components are given by (remembering (3.16) and (3.17) with $dl''=0$)

$$\begin{array}{lll} dl_1=dx_1 & dl_2=\varepsilon_{21}\,dx_1 & dl_3=0 \\ d\bar{l}_1=\varepsilon_{12}\,dx_2 & d\bar{l}_2=dx_2 & d\bar{l}_3=0. \end{array} \tag{3.36}$$

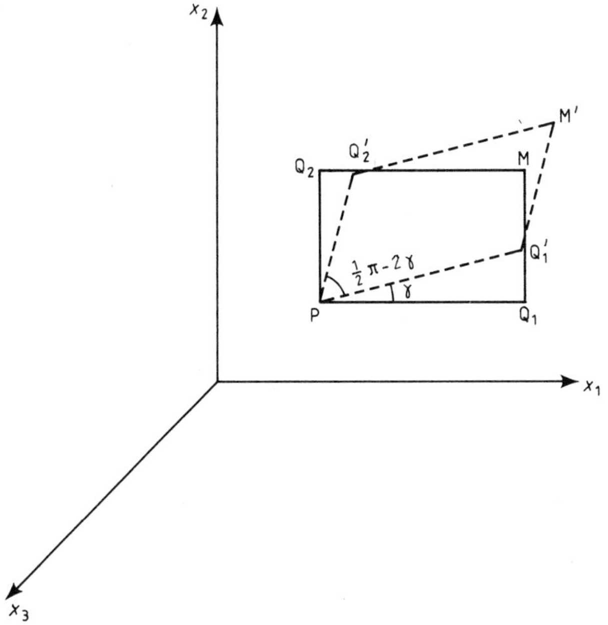

Figure 3.4 Diagram showing the displacements of the line elements PQ_1 and PQ_2 and the deformation of the rectangle PQ_1MQ_2 due to the pure shear ε_{12} ($=\gamma$).

Hence the angle θ between the two lines PQ_1' and PQ_2' is given by

$$\cos\theta = 2\varepsilon_{12} \tag{3.37}$$

where we have used the fact that $\varepsilon_{12}=\varepsilon_{21}$ and, as usual, ignored squares and higher powers of ε_{12}. If we denote by 2γ the change in the angle between the two line elements originally perpendicular to each other and parallel to the x_1- and x_2-axes, then $\theta=\frac{1}{2}\pi-2\gamma$ and since $\sin 2\gamma \simeq 2\gamma$, we have

$$\gamma=\varepsilon_{12} \tag{3.38}$$

which is the interpretation of ε_{12} that we were seeking. (Note that, to linear terms in ε_{ij}, γ is given by (3.38) even if other $\varepsilon_{ij}\neq 0$.) The physical significances of ε_{13} and ε_{23} are similar and refer, respectively to the x_1–$x_3$and x_2–x_3 elements.

Returning to ε_{12}, figure 3.4 depicts the positions of PQ_1, PQ_1' etc as well as the deformation of the rectangle PQ_1MQ_2 due to ε_{12}. (This may be contrasted with figure 3.3 which depicts the effect of a rotation about the x_3-axis.) A cylindrical body with rectangular cross section perpendicular to the x_3-axis is deformed due to ε_{12} such that each of these cross sections is a parallelogram $PQ_1'M'Q_2'$. Such a deformation is called pure shear (in the x_1–x_2 plane).

We observe that since $\Delta=\varepsilon_{ii}=0$, no change in volume occurs and a pure shear is isovoluminous. Finally, we remark that in physical problems one sometimes encounters what is known as a simple shear. A simple shear of angle

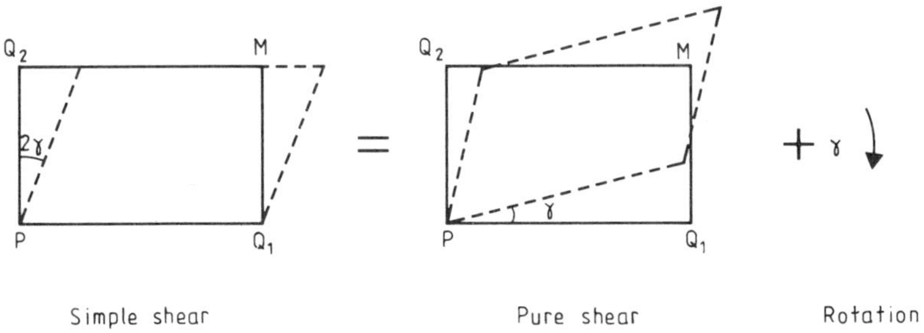

Figure 3.5 This diagram shows a simple shear and its equivalence to the sum of a pure shear and a rotation.

2γ in the x_1–x_2 plane is depicted in figure 3.5 which also illustrates how a simple shear is a sum of a pure shear and a rotation. One readily verifies that the strain involved is the same in the simple- and pure-shear cases and the displacements in the two differ only by a rotation. We note that for a homogeneous deformation the displacements in the pure-shear case (in the x_1–x_2 plane) are

$$s_1 = \gamma x_2 \qquad s_2 = \gamma x_1 \qquad s_3 = 0 \tag{3.39}$$

and in simple shear

$$s_1 = 2\gamma x_2 \qquad s_2 = 0 \qquad s_3 = 0. \tag{3.40}$$

3.4 Strain tensors for finite deformations

Although, as mentioned in §3.2.1, the subject of elasticity for finite deformations lies outside the scope of the present work, it is instructive to consider briefly the problem of defining the strain tensor for this case also.

For finite deformations, that is, when the squares and products of $\partial s_i/\partial x_j$ cannot be neglected compared with the linear terms $\partial s_i/\partial x_j$, the strain can be essentially specified by two different strain tensors. For reasons which will become clear presently they are usually referred to as the Lagrangian and Eulerian strain tensors. We consider the former case first.

3.4.1 The Lagrangian strain tensor

We recall from §3.2.1 and figure 3.2 that an infinitesimal line element $d\boldsymbol{x}$ joining a particle at $\boldsymbol{x}$ and a particle at $\boldsymbol{x}+d\boldsymbol{x}$ before deformation, occupies in the deformed specimen the position P′Q′ with the vector $d\boldsymbol{l}$ joining the same two particles having the components given by (3.10). Rewriting (3.10) with an obvious relabelling of suffixes, we have for the kth component:

$$\mathrm{d}l_k = \left(\delta_{ki} + \frac{\partial s_k}{\partial x_i}\right) \mathrm{d}x_i. \tag{3.41}$$

The square of the length of the vector $\mathrm{d}\boldsymbol{l}$ is

$$\begin{aligned}|\mathrm{d}\boldsymbol{l}|^2 &= \sum_k (\mathrm{d}l_k)^2 = \sum_k \sum_i \sum_j \left(\delta_{ki} + \frac{\partial s_k}{\partial x_i}\right)\left(\delta_{kj} + \frac{\partial s_k}{\partial x_j}\right) \mathrm{d}x_i\, \mathrm{d}x_j \\ &= \sum_{i,j} \mathrm{d}x_i\, \mathrm{d}x_j \sum_k \left(\delta_{ki}\delta_{kj} + \delta_{ki}\frac{\partial s_k}{\partial x_j} + \delta_{kj}\frac{\partial s_k}{\partial x_i} + \frac{\partial s_k}{\partial x_i}\frac{\partial s_k}{\partial x_j}\right) \\ &= \sum_{i,j} \mathrm{d}x_i\, \mathrm{d}x_j \left(\delta_{ij} + \frac{\partial s_i}{\partial x_j} + \frac{\partial s_j}{\partial x_i} + \sum_k \frac{\partial s_k}{\partial x_i}\frac{\partial s_k}{\partial x_j}\right).\end{aligned} \tag{3.42}$$

Hence we may write

$$|\mathrm{d}\boldsymbol{l}|^2 = |\mathrm{d}\boldsymbol{x}|^2 + 2E_{ij}\, \mathrm{d}x_i\, \mathrm{d}x_j \tag{3.43}$$

where

$$E_{ij} = E_{ji} = \frac{1}{2}\left(\frac{\partial s_i}{\partial x_j} + \frac{\partial s_j}{\partial x_i}\right) + \frac{1}{2}\frac{\partial s_k}{\partial x_i}\frac{\partial s_k}{\partial x_j} \tag{3.44}$$

and where in writing (3.43) and (3.44) we have reverted to using the Einstein summation convention.

We observe from (3.43) that if all $E_{ij} \equiv 0$, then $|\mathrm{d}\boldsymbol{l}| = |\mathrm{d}\boldsymbol{x}|$ for every line element $\mathrm{d}\boldsymbol{x}$. Hence the E_{ij} specify the strain in the region around $\boldsymbol{x}$. For small deformations the second term in (3.44) can be neglected compared with the first and then $E_{ij} = \varepsilon_{ij}$, the strain tensor introduced in equation (3.12) for infinitesimal deformations. The second term in (3.44) may be verified to transform like the components of a second-rank tensor so that E_{ij} itself is a second-rank tensor. E_{ij} is often referred to as the Lagrangian strain tensor. The specification of the rotations in terms of $\partial s_i/\partial x_j$ for finite deformations is involved and will not be discussed here (see, for example, Love 1944).

The physical significance of the individual terms E_{ij} is similar to that for the corresponding ε_{ij} except that one must not now regard E_{ij} or $\partial s_i/\partial x_j$ as small. Thus the length $|\mathrm{d}\boldsymbol{l}|$ of a long thin filament lying along the x_1-axis and of length $\mathrm{d}x_1$ before deformation is given by, from (3.43):

$$|\mathrm{d}\boldsymbol{l}|^2 = (\mathrm{d}x_1)^2 + 2E_{11}(\mathrm{d}x_1)^2 = (1 + 2E_{11})(\mathrm{d}x_1)^2. \tag{3.45}$$

Hence the extension, per unit length, of the fibre is

$$(|\mathrm{d}\boldsymbol{l}| - \mathrm{d}x_1)/\mathrm{d}x_1 = (1 + 2E_{11})^{1/2} - 1. \tag{3.46}$$

For infinitesimal deformations $E_{11} \simeq \varepsilon_{11}$ and $\varepsilon_{11} \ll 1$, so that the right-hand side of (3.46) reduces to just ε_{11} in agreement with the expression (3.33) for the extension given earlier.

Next to interpret E_{12}, again consider two line elements PQ_1 and PQ_2 parallel to the x_1- and x_2-axes and of lengths dx_1 and dx_2 respectively as in figure 3.4. Denoting the two vectors after deformation by $d\boldsymbol{l}$ and $d\bar{\boldsymbol{l}}$ respectively, their components, using (3.41), are

$$dl_1=\left(1+\frac{\partial s_1}{\partial x_1}\right)dx_1 \qquad dl_2=\frac{\partial s_2}{\partial x_1}dx_1 \qquad dl_3=\frac{\partial s_3}{\partial x_1}dx_1 \tag{3.47a}$$

$$d\bar{l}_1=\frac{\partial s_1}{\partial x_2}dx_2 \qquad d\bar{l}_2=\left(1+\frac{\partial s_2}{\partial x_2}\right)dx_2 \qquad d\bar{l}_3=\frac{\partial s_3}{\partial x_2}dx_2. \tag{3.47b}$$

Hence the angle θ between $d\boldsymbol{l}$ and $d\bar{\boldsymbol{l}}$ may be seen to be given by

$$d\boldsymbol{l}\cdot d\bar{\boldsymbol{l}}=|d\boldsymbol{l}|\,|d\bar{\boldsymbol{l}}|\cos\theta=2E_{12}\,dx_1\,dx_2. \tag{3.48}$$

However, from (3.46), $|d\boldsymbol{l}|=(1+2E_{11})^{1/2}\,dx_1$ and $|d\bar{\boldsymbol{l}}|=(1+2E_{22})^{1/2}\,dx_2$. Hence

$$\cos\theta=\frac{2E_{12}}{(1+2E_{11})^{1/2}(1+2E_{22})^{1/2}} \tag{3.49}$$

which again reduces to the expression (3.37) for small deformations.

3.4.2 The Eulerian strain tensor

For finite deformations one sometimes uses another expression for the strain tensor. We recall that in our notation the position of a particle before deformation is denoted by $\boldsymbol{x}$, and its position after a displacement $\boldsymbol{s}$ by $\boldsymbol{l}=\boldsymbol{x}+\boldsymbol{s}$. In equation (3.7) and subsequent work we regarded $\boldsymbol{s}$ and therefore $\boldsymbol{l}$ as a function of the undisplaced position ($\boldsymbol{x}$) of the particle. In mechanics and hydrodynamics this choice of independent variables is referred to as the Lagrangian scheme and hence the usage of the name Lagrangian strain tensor for E_{ij}. If, on the other hand, we take the displacement $\boldsymbol{s}$, and therefore also $\boldsymbol{x}$, as a function of the displaced position ($\boldsymbol{l}$) of the particle, then we have essentially the Eulerian scheme. To derive the expression for the corresponding strain tensor, we first rewrite the relation (3.7) between $\boldsymbol{x}$, $\boldsymbol{s}$ and $\boldsymbol{l}$ in the form

$$x(\boldsymbol{l})=\boldsymbol{l}-s(\boldsymbol{l}) \qquad \boldsymbol{x}(\boldsymbol{l}+d\boldsymbol{l})=\boldsymbol{l}+d\boldsymbol{l}-\boldsymbol{s}(\boldsymbol{l}+d\boldsymbol{l}). \tag{3.50}$$

Just as (3.7) leads to (3.10) or (3.41) for the components dl_k of $d\boldsymbol{l}$ in terms of dx_i, so also (3.50) gives expressions for the components dx_k of $d\boldsymbol{x}$ in the form:

$$dx_k=\left(\delta_{ki}-\frac{\partial s_k}{\partial l_i}\right)dl_i. \tag{3.51}$$

Forming $|d\boldsymbol{x}|^2$ from (3.51) one obtains straightforwardly

$$|d\boldsymbol{x}|^2=|d\boldsymbol{l}|^2-2\mathscr{E}_{ij}\,dl_i\,dl_j$$

or

$$|\mathrm{d}\boldsymbol{l}|^2 = |\mathrm{d}\boldsymbol{x}|^2 + 2\mathscr{E}_{ij}\,\mathrm{d}l_i\,\mathrm{d}l_j \tag{3.52}$$

where

$$\mathscr{E}_{ij} = \mathscr{E}_{ji} = \frac{1}{2}\left(\frac{\partial s_i}{\partial l_j} + \frac{\partial s_j}{\partial l_i}\right) - \frac{1}{2}\frac{\partial s_k}{\partial l_i}\frac{\partial s_k}{\partial l_j} \tag{3.53}$$

is the Eulerian strain tensor.

We note that in the Eulerian strain tensor the derivatives of $\boldsymbol{s}$ are evaluated at the displaced position ($\boldsymbol{l}$) of the particle, whereas in the Lagrangian strain tensor they are evaluated as the undisplaced position of the particle. The two tensors are, in general, different from one another although, of course, they are related to each other by virtue of the fact that (3.7) and (3.50) are not independent of one another. For small deformations $\mathscr{E}_{ij} \simeq E_{ij}$, we have, remembering (3.51),

$$\frac{\partial s_i}{\partial l_j} = \frac{\partial s_i}{\partial x_k}\frac{\partial x_k}{\partial l_j} = \frac{\partial s_i}{\partial x_k}\left(\delta_{kj} - \frac{\partial s_k}{\partial l_j}\right) \simeq \frac{\partial s_i}{\partial x_k} \tag{3.54}$$

on neglecting squares and products of $\partial s_i/\partial x_j$ or $\partial s_i/\partial l_j$. Hence in the limit of infinitesimal deformations $\mathscr{E}_{ij} = E_{ij} = \varepsilon_{ij}$.

For a further discussion of this topic, as well as other topics on finite deformations, see the references given in §3.2.1. We shall now follow our usual course of discussion on infinitesimal strains.

3.5 Compatibility conditions for infinitesimal strains

Since all the strain components, ε_{ij}, as defined in equation (3.12), are obtained by differentiating the displacement $\boldsymbol{s}$ with respect to x_1, x_2 and x_3, they are not independent. That is to say the functions ε_{ij} cannot be arbitrarily assigned but instead they satisfy certain relations. These relations are called compatibility conditions and are useful in many ways as, for example, in checking the solutions of equilibrium problems.

We can arrive at these conditions simply by using the expressions

$$\varepsilon_{11} = \frac{\partial s_1}{\partial x_1} \qquad \varepsilon_{22} = \frac{\partial s_2}{\partial x_2} \qquad \text{and } \varepsilon_{33} = \frac{\partial s_3}{\partial x_3} \tag{3.55}$$

which in turn allow us to write

$$\frac{\partial^2 \varepsilon_{11}}{\partial x_2^2} + \frac{\partial^2 \varepsilon_{22}}{\partial x_1^2} = 2\frac{\partial^2 \varepsilon_{12}}{\partial x_1\,\partial x_2} \tag{3.56}$$

$$\frac{\partial^2 \varepsilon_{22}}{\partial x_3^2} + \frac{\partial^2 \varepsilon_{33}}{\partial x_2^2} = 2\frac{\partial^2 \varepsilon_{23}}{\partial x_2\,\partial x_3} \tag{3.57}$$

$$\frac{\partial^2 \varepsilon_{11}}{\partial x_3^2}+\frac{\partial^2 \varepsilon_{33}}{\partial x_1^2}=2\frac{\partial^2 \varepsilon_{13}}{\partial x_1\,\partial x_3} \tag{3.58}$$

$$\frac{\partial^2 \varepsilon_{11}}{\partial x_2\,\partial x_3}=\frac{\partial}{\partial x_1}\left(-\frac{\partial \varepsilon_{23}}{\partial x_1}+\frac{\partial \varepsilon_{31}}{\partial x_2}+\frac{\partial \varepsilon_{12}}{\partial x_3}\right) \tag{3.59}$$

$$\frac{\partial^2 \varepsilon_{22}}{\partial x_3\,\partial x_1}=\frac{\partial}{\partial x_2}\left(\frac{\partial \varepsilon_{23}}{\partial x_1}-\frac{\partial \varepsilon_{31}}{\partial x_2}+\frac{\partial \varepsilon_{12}}{\partial x_3}\right) \tag{3.60}$$

$$\frac{\partial^2 \varepsilon_{33}}{\partial x_1\,\partial x_2}=\frac{\partial}{\partial x_3}\left(\frac{\partial \varepsilon_{23}}{\partial x_1}+\frac{\partial \varepsilon_{31}}{\partial x_2}-\frac{\partial \varepsilon_{12}}{\partial x_3}\right). \tag{3.61}$$

The above six relations (3.56) through (3.61) are known as the compatibility conditions for small strains.

3.6 Tensor and strain tensors in cylindrical and spherical polar coordinate systems

3.6.1 Components of a tensor of rank two in cylindrical and spherical polar coordinates

In Chapter 2 tensors were treated in a Cartesian coordinate system. However, if a given problem has some inherent symmetries then it becomes easier to work in the cylindrical or spherical polar coordinate system (see, for example, §5.2). The purpose of this section is to derive tensor components in these two coordinate systems. Amongst others† one simple way of doing this is to employ the method of transformation of coordinates. We recall the transformation equations for a tensor of rank two (see equation (2.28))

$$T'_{ij}=a_{im}a_{jn}T_{mn}$$

where the prime indicates components in the desired coordinate system and the unprimed components refer to the reference coordinates.

(*a*) *Cylindrical coordinates* (r, φ, z)

Consider a point P at a distance r from the origin of the Cartesian reference axes x_1, x_2 and x_3 (see figure 3.6). Now at P we consider a new frame of rectangular coordinates x'_1, x'_2 and x'_3 such that they correspond to cylindrical coordinates r, φ, z respectively, i.e.

$$x'_1 \mathrel{\hat{=}} r \qquad x'_2 \mathrel{\hat{=}} \varphi \qquad \text{and } x'_3\equiv x_3\equiv z \tag{3.62}$$

where $\hat{=}$ stands for 'corresponds to'.

† Actually the problem of treating tensor components in a curvilinear coordinate system is quite involved and can be found in any advanced book on tensor analysis. Here we consider only a simplistic approach to the problem.

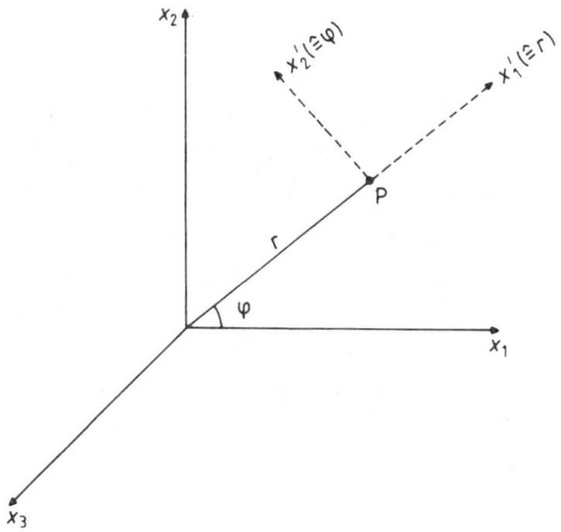

Figure 3.6 This diagram shows the transformation from Cartesian coordinates (x_1, x_2, x_3) to cylindrical coordinates (r, φ, z).

Note that for different positions of P the directions of the x_1'-, x_2'- and x_3'-axes are also different. By inspection one sees that the x_1'-, x_2'- and x_3'-axes can be obtained from x_1, x_2 and x_3 by rotation through an angle φ about the x_3-axis. The transformation matrix (see (P2.1)) is:

$$\mathbf{a}=\begin{pmatrix} \cos\varphi & \sin\varphi & 0 \\ -\sin\varphi & \cos\varphi & 0 \\ 0 & 0 & 1 \end{pmatrix}. \tag{3.63}$$

Employing this in equation (2.28) the components T_{ij}' in cylindrical polar coordinates become

$$T_{11}' \equiv T_{rr} = \cos^2\varphi\, T_{11} + \cos\varphi\sin\varphi(T_{12}+T_{21}) + \sin^2\varphi\, T_{22} \tag{3.64}$$

$$T_{12}' \equiv T_{r\varphi} = -\cos\varphi\sin\varphi\, T_{11} + \cos^2\varphi\, T_{12} - \sin^2\varphi\, T_{21} + \sin\varphi\cos\varphi\, T_{22} \tag{3.65}$$

$$T_{21}' \equiv T_{\varphi r} = -\cos\varphi\sin\varphi T_{11} - \sin^2\varphi T_{12} + \cos^2\varphi T_{21} + \cos\varphi\sin\varphi\, T_{22} \tag{3.66}$$

$$T_{22}' \equiv T_{\varphi\varphi} = \sin^2\varphi\, T_{11} - \sin\varphi\cos\varphi\, T_{12} - \sin\varphi\cos\varphi\, T_{21} + \cos^2\varphi\, T_{22} \tag{3.67}$$

$$T_{13}' \equiv T_{rz} = \cos\varphi\, T_{13} + \sin\varphi\, T_{23} \tag{3.68}$$

$$T_{31}' \equiv T_{zr} = \cos\varphi\, T_{31} + \sin\varphi\, T_{32} \tag{3.69}$$

$$T_{23}' \equiv T_{\varphi z} = -\sin\varphi\, T_{13} + \cos\varphi\, T_{23} \tag{3.70}$$

$$T'_{32} \equiv T_{z\varphi} = -\sin\varphi\, T_{31} + \cos\varphi\, T_{32} \tag{3.71}$$

$$T'_{33} \equiv T_{zz} = T_{33}. \tag{3.72}$$

Thus there are in total nine independent components. It should be noted that the components T_{11}, T_{12}, etc appearing on the right-hand side of expressions (3.64) to (3.72) have to be expressed explicitly in terms of cylindrical coordinates as is shown in §3.6.2. If $\mathbf{T}$ is a symmetric tensor then $T_{ij} = T_{ji}$, hence substituting $T_{12} = T_{21}$, $T_{13} = T_{31}$ and $T_{23} = T_{32}$ in equations (3.64) to (3.72), one obtains six independent components, namely T_{rr}, $T_{r\varphi} = T_{\varphi r}$, $T_{\varphi\varphi}$, $T_{rz} = T_{zr}$, $T_{\varphi z} = T_{z\varphi}$ and T_{zz}. Similarly, for antisymmetric tensors the components are obtained by putting $T_{ij} = -T_{ij}$ in equations (3.64) to (3.72).

(*b*) *Spherical polar coordinates* (r, θ, φ)

Following the same approach as in (*a*) we set up primed axes x'_1, x'_2, x'_3 at the point P such that

$$x'_1 \mathrel{\hat{=}} \theta \qquad x'_2 \mathrel{\hat{=}} \varphi \qquad \text{and } x'_3 \mathrel{\hat{=}} r. \tag{3.73}$$

The transformation (3.73) can be achieved using Cartesian axes x_1, x_2, x_3 in two successive operations: (i) rotate the axes x_1, x_2, x_3 about the x_3-axis through an angle φ giving new axes x''_1, x''_2 and $x''_3 = x_3$, (ii) then rotate through an angle θ about x''_2 giving single primed axes x'_1, $x'_2 = x''_2$, x'_3.

The transformation equation for operation (i) is

$$\boldsymbol{x}'' = \mathbf{a}_1 \boldsymbol{x} \tag{3.74}$$

where $\mathbf{a}_1$ is the same as in (P2.1).

The transformation equation for the operation (ii) is

$$\boldsymbol{x}' = \mathbf{a}_2 \boldsymbol{x}'' \tag{3.75}$$

where

$$\mathbf{a}_2 = \begin{pmatrix} \cos\theta & 0 & -\sin\theta \\ 0 & 1 & 0 \\ \sin\theta & 0 & \cos\theta \end{pmatrix}. \tag{3.76}$$

From relations (3.74) and (3.75) one has

$$\boldsymbol{x}' = \mathbf{a}_2 \mathbf{a}_1 \boldsymbol{x} \equiv \mathbf{a} \boldsymbol{x} \tag{3.77}$$

where

$$\mathbf{a} = \begin{pmatrix} \cos\theta\cos\varphi & \cos\theta\sin\varphi & -\sin\theta \\ -\sin\varphi & \cos\varphi & 0 \\ \sin\theta\cos\varphi & \sin\theta\sin\varphi & \cos\theta \end{pmatrix}. \tag{3.78}$$

Substituting this in equation (2.28) one obtains components of $\mathbf{T}$ in spherical polar coordinates.

3.6.2 Components of strain and rotation tensors in cylindrical coordinates

Let s_r, s_φ, s_z represent displacements in a cylindrical coordinate system which correspond to the displacements s_1, s_2, s_3 in Cartesian coordinates. Using the transformation matrix (3.63) one writes

$$s_1 = s_r \cos\varphi - s_\varphi \sin\varphi \tag{3.79a}$$

$$s_2 = s_r \sin\varphi + s_\varphi \cos\varphi \tag{3.79b}$$

$$s_3 = s_z. \tag{3.79c}$$

We also have relations between the Cartesian coordinates x_1, x_2, x_3 and the cylindrical coordinates r, φ, z:

$$x_1 = r\cos\varphi \qquad x_2 = r\sin\varphi \qquad x_3 = z \tag{3.80}$$

which allow us to write

$$\frac{\partial}{\partial x_1} = \frac{\partial r}{\partial x_1}\frac{\partial}{\partial r} + \frac{\partial\varphi}{\partial x_1}\frac{\partial}{\partial\varphi} = \cos\varphi\frac{\partial}{\partial r} - \frac{\sin\varphi}{r}\frac{\partial}{\partial\varphi} \tag{3.81a}$$

$$\frac{\partial}{\partial x_2} = \frac{\partial r}{\partial x_2}\frac{\partial}{\partial r} + \frac{\partial\varphi}{\partial x_2}\frac{\partial}{\partial\varphi} = \sin\varphi\frac{\partial}{\partial r} + \frac{\cos\varphi}{r}\frac{\partial}{\partial\varphi} \tag{3.81b}$$

$$\frac{\partial}{\partial x_3} = \frac{\partial}{\partial z}. \tag{3.81c}$$

Equations (3.79) and (3.81) are now used in (3.12) to obtain the strain components ε_{11}, ε_{12} ... etc in terms of cylindrical coordinates. In turn when these are submitted in equations (3.64) to (3.72) (replace **T** by ε), we obtain

$$\varepsilon_{rr} = \partial s_r/\partial r \tag{3.82}$$

$$\varepsilon_{\varphi\varphi} = \frac{s_r}{r} + \frac{1}{r}\frac{\partial s_\varphi}{\partial\varphi} \tag{3.83}$$

$$\varepsilon_{zz} = \partial s_z/\partial z \tag{3.84}$$

$$\varepsilon_{r\varphi} = \frac{1}{2}\left(\frac{1}{r}\frac{\partial s_r}{\partial\varphi} + \frac{\partial s_\varphi}{\partial r} - \frac{s_\varphi}{r}\right) \tag{3.85}$$

$$\varepsilon_{zr} = \frac{1}{2}\left(\frac{\partial s_r}{\partial z} + \frac{\partial s_z}{\partial r}\right) \tag{3.86}$$

$$\varepsilon_{z\varphi} = \frac{1}{2}\left(\frac{1}{r}\frac{\partial s_z}{\partial\varphi} + \frac{\partial s_\varphi}{\partial z}\right) \tag{3.87}$$

the components of strain tensor in cylindrical coordinates. The corresponding dilatation, Δ, is:

$$\Delta=\varepsilon_{rr}+\varepsilon_{\varphi\varphi}+\varepsilon_{zz}$$

$$=\frac{1}{r}\frac{\partial}{\partial r}(rs_r)+\frac{1}{r}\frac{\partial s_\varphi}{\partial \varphi}+\frac{\partial s_z}{\partial z}. \tag{3.88}$$

Similarly, if R_r, R_φ and R_z denote the components of rotation in cylindrical coordinates then remembering that $\boldsymbol{R}=\frac{1}{2}\operatorname{curl}\boldsymbol{s}$, one can show that

$$2R_r=\frac{1}{r}\frac{\partial s_z}{\partial \varphi}-\frac{\partial s_\varphi}{\partial z} \tag{3.89}$$

$$2R_\varphi=\frac{\partial s_r}{\partial z}-\frac{\partial s_z}{\partial r} \tag{3.90}$$

$$2R_z=\frac{1}{r}\frac{\partial}{\partial r}(rs_\varphi)-\frac{1}{r}\frac{\partial s_r}{\partial \varphi}. \tag{3.91}$$

These results are of fundamental importance in the theory of vorticity, which we will pursue later on.

3.7 Thermal-expansion coefficients

Before closing our general discussion on second-rank tensors we briefly discuss an observable, the thermal-expansion coefficient, one of the intrinsic second-rank tensor properties of a crystal. The coefficients of thermal expansion α_{ij} measured at constant stress† σ are defined as

$$\alpha_{ij}=\left(\frac{\partial \varepsilon_{ij}}{\partial T}\right)_\sigma. \tag{3.92}$$

Like ε_{ij}, the elements α_{ij} are also the components

$$\alpha_{ij}=\begin{pmatrix}\alpha_{11} & \alpha_{12} & \alpha_{13}\\ \alpha_{21} & \alpha_{22} & \alpha_{23}\\ \alpha_{31} & \alpha_{32} & \alpha_{33}\end{pmatrix} \tag{3.93}$$

of a second-rank tensor. Since the temperature, T, in equation (3.92) is a scalar quantity the resulting ε_{ij} must comply with the symmetry of the crystal and so also α_{ij}. Our basic objective here is to obtain the components α_{ij} conforming to some of the commonly used crystal symmetry classes‡.

† Stress will be discussed in Chapter 4.
‡ Readers may familiarise themselves with crystal symmetry from, for exampie, Phillips (1971).

(a) Monoclinic system

The monoclinic system possesses one twofold axis of symmetry. It remains invariant against a rotation of 180° about the axis of symmetry, say the x_3-axis. From the transformation of coordinates ($x'_1 = -x_1$, $x'_2 = -x_2$, $x'_3 = x_3$, see equation (2.10)),

$$\mathbf{a} = \begin{pmatrix} -1 & 0 & 0 \\ 0 & -1 & 0 \\ 0 & 0 & 1 \end{pmatrix}.$$

The thermal-expansion coefficients $\boldsymbol{\alpha}'$ referred to the primed axes are:

$$\boldsymbol{\alpha}' = \mathbf{a}\boldsymbol{\alpha}\mathbf{a}^{-1} = \begin{pmatrix} -1 & 0 & 0 \\ 0 & -1 & 0 \\ 0 & 0 & 1 \end{pmatrix} \begin{pmatrix} \alpha_{11} & \alpha_{12} & \alpha_{13} \\ \alpha_{21} & \alpha_{22} & \alpha_{23} \\ \alpha_{31} & \alpha_{32} & \alpha_{33} \end{pmatrix} \begin{pmatrix} -1 & 0 & 0 \\ 0 & -1 & 0 \\ 0 & 0 & 1 \end{pmatrix}. \tag{3.94}$$

Using the rules of matrix multiplication (see, for example, Aitken (1956)) one obtains

$$\boldsymbol{\alpha}' = \begin{pmatrix} \alpha_{11} & \alpha_{12} & -\alpha_{13} \\ \alpha_{21} & \alpha_{22} & -\alpha_{23} \\ -\alpha_{31} & -\alpha_{32} & +\alpha_{33} \end{pmatrix}. \tag{3.95}$$

Now since $\boldsymbol{\alpha}' = \boldsymbol{\alpha}$, we therefore must have

$$\alpha_{13} = \alpha_{23} = \alpha_{31} = \alpha_{32} = 0. \tag{3.96}$$

For a monoclinic system $\boldsymbol{\alpha}$ is then:

$$\boldsymbol{\alpha} = \begin{pmatrix} \alpha_{11} & \alpha_{12} & 0 \\ \alpha_{21} & \alpha_{22} & 0 \\ 0 & 0 & \alpha_{33} \end{pmatrix} \tag{3.97}$$

Thus we have only four independent components of $\boldsymbol{\alpha}$, i.e. $\alpha_{11}, \alpha_{22}, \alpha_{33}$ and α_{12} ($\equiv \alpha_{21}$).

(b) Orthorhombic system

This system possesses three mutually perpendicular twofold axes of symmetry. The rule which is true for the x_3-axis of symmetry for a monoclinic system is now also true for the x_1- and x_2-axes of symmetry. This symmetry, in addition to (3.97), requires that $\alpha_{12} = \alpha_{21} = 0$ and hence for the orthorhombic system:

$$\boldsymbol{\alpha} = \begin{pmatrix} \alpha_{11} & 0 & 0 \\ 0 & \alpha_{22} & 0 \\ 0 & 0 & \alpha_{33} \end{pmatrix}. \tag{3.98}$$

(c) Cubic system

In a cubic system the x_1-, x_2-, x_3-axes cannot be distinguished from one another and each of them represent fourfold axes of symmetry. If we carry out an anticlockwise rotation of 90° about the x_3-axis ($x'_1=x_2, x'_2=-x_1, x'_3=x_3$) the invariant condition $\boldsymbol{\alpha}'=\boldsymbol{\alpha}$ requires that $\alpha_{11}=\alpha_{22}$ and $\alpha_{ij}=0$ $(i\neq j)$. Similarly, a rotation of 90° about the x_1-axis ($x'_1=x_1$, $x'_2=x_3$, $x'_3=-x_2$) implies that $\alpha_{22}=\alpha_{33}$, $\alpha_{ij}=0$ $(i\neq j)$. Thus, in a cubic system, the thermal expansions in all directions are the same ($\alpha_{11}=\alpha_{22}=\alpha_{33}$). $\boldsymbol{\alpha}$ is evidently given by

$$\boldsymbol{\alpha}=\begin{pmatrix}\alpha_{11} & 0 & 0\\ 0 & \alpha_{11} & 0\\ 0 & 0 & \alpha_{11}\end{pmatrix}. \tag{3.99}$$

(d) Hexagonal system

Here there is one sixfold axis of symmetry. The transformation matrix for a rotation of 60° about the x_3-axis is

$$\mathbf{a}=\begin{pmatrix}\frac{1}{2} & \sqrt{3}/2 & 0\\ -\sqrt{3}/2 & \frac{1}{2} & 0\\ 0 & 0 & 1\end{pmatrix}. \tag{3.100}$$

In conjunction with the invariant condition this requires that $\alpha_{11}=\alpha_{22}$ and $\alpha_{ij}=0$ $(i\neq j)$. Hence $\boldsymbol{\alpha}$ for the hexagonal system becomes

$$\boldsymbol{\alpha}=\begin{pmatrix}\alpha_{11} & 0 & 0\\ 0 & \alpha_{11} & 0\\ 0 & 0 & \alpha_{33}\end{pmatrix}. \tag{3.101}$$

Finally, it may be mentioned that the present method of treating crystal symmetry is also applicable to some other intrinsic tensor properties of rank two, for example, electrical conductivity, permittivity, permeability, etc. For a more detailed treatment Nye (1957), Bhagavantam (1966) and Wooster (1973) may be consulted.

Problems

3.1 The displacement vector $\boldsymbol{s}$ in two deformations is given by (with constants k_1 and k_2)

$$\text{(i)}\quad s_1=k_1x_2,\quad s_2=k_2x_1,\quad s_3=0 \qquad \text{(ii)}\quad s_1=\tfrac{1}{2}(k_1+k_2)x_2,\quad s_2=\tfrac{1}{2}(k_1+k_2)x_1,\quad s_3=0. \tag{P3.1}$$

Determine the strain tensor, its principal values and directions of principal values for the two cases.

The two displacements (i) and (ii) are obviously not identical but the quantities calculated are the same for the two cases. Why is this?

3.2 In a given problem all other strain components are zero except

$$\varepsilon_{11}=k_1(x_1^2-x_2^2) \qquad \varepsilon_{22}=k_1x_1x_2 \qquad \varepsilon_{12}=k_2x_1x_2 \qquad \text{(P3.2)}$$

where $k_1=3\times10^{-5}$ and $k_2=3.5\times10^{-5}$. Is the above strain compatible? If not, for what values of k_1 and k_2 would the strain be compatible?

3.3 In a homogeneous deformation of a material if the principal strains are

$$\varepsilon^{(1)}=3\times10^{-5} \qquad \varepsilon^{(2)}=3\times10^{-5} \qquad \varepsilon^{(3)}=-6\times10^{-5}$$

then consider:

(i) A long straight filament (with negligible transverse dimensions) along the direction whose polar coordinates are (θ, φ) with x_3 as polar axis. If before the deformation the length of the filament is L, what is its length (L') after the deformation? Sketch the variation of $L'-L$ with θ and φ.

(ii) A small element in the material before the deformation has a volume 10^{-6} cm^3. What is the volume after deformation?

3.4 A long solid rod of circular cross section undergoes the following displacements (a, b, c, e are constants):

$$s_r=ar+br^{-1} \qquad s_\varphi=crz \qquad s_z=ez$$

where (r, φ, z) are cylindrical polar coordinates with z along the axis of the cylinder. Determine the expressions for the components of the strain tensor, the rotation tensor and for the dilatation.

3.5 Obtain general expressions for the components of the strain tensor, the rotation tensor and for the dilatation in spherical polar coordinates (r, θ, φ).

3.6 The thermal-expansion tensor of a hexagonal crystal referred to its principal axes is

$$\alpha_1=\alpha_2=5\times10^{-5} \qquad \alpha_3=8\times10^{-5}\text{ in K}^{-1}.$$

Consider a long straight filament (with negligible transverse dimensions) along the direction whose polar coordinates are (θ, φ) with the polar axis as the x_3-axis then (i) calculate by what fraction the length of the filament increased when it is heated from 300 to 310 K? (ii) Obtain a numerical value for this fractional increase if $\theta=30°$ and $\varphi=47°$.

4

STRESS TENSOR AND ELASTIC CONSTANTS

The concept of stress in continuum mechanics is very useful in expressing the interaction between one part of a material body and another. When a material body is in contact with another body, or two parts of the same material body are in contact through an imaginary surface, then there are always material actions and reactions across the surface. The matter to the right exerts a force on the matter to the left and *vice versa*. The force acting per unit area is called stress. If the force acts outwards in the direction normal to the cross sectional area, it is called tensile stress but if it is directed inwards along the normal direction, it is called thrust. Stress is positive if it is tensile and negative if it is compressive. Unlike fluids at rest, the forces acting on any given plane face in solids are not necessarily normal to the face. This is because solids possess rigidity. If we wish to calculate the stress at a point in a continuum solid we need to know the force acting on every unit area that passes through that point. It is sufficient to consider three such unit areas at a point, each representing a plane in the Cartesian sense. Since one has three components of stress for each plane, there are, in general, nine components of stress at a point in a solid. The present Chapter is concerned with the specification of stress components followed by stress–strain relations for elastic material bodies (especially isotropic solids).

PART I. STRESS COMPONENTS AND SOME BASIC EQUATIONS OF ELASTICITY

4.1 The stress components

Consider the forces acting on an element of volume in the form of a rectangular parallelepiped (see figure 4.1) due to the reaction of matter surrounding it. The force acting on a given face is proportional to the area. Let us denote the components of the force per unit area acting on a face perpendicular to the x_1-axis by σ_{11}, σ_{12} and σ_{13} as shown in figure 4.1. Similarly, for faces perpendicular to the x_2- and x_3-axes let the components be $\sigma_{22}, \sigma_{23}, \sigma_{21}$ and $\sigma_{33}, \sigma_{31}, \sigma_{32}$ respectively. The first suffix on σ, in our notation, denotes the

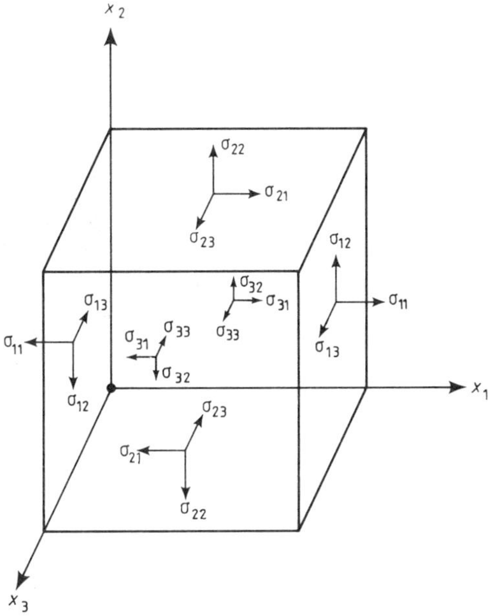

Figure 4.1 The components of stress on the faces of a rectangular parallelepiped.

plane and the second denotes the component direction of the force†. Regarding the direction of these components, σ_{11} is positive if it is an extensional force, i.e. directed along the positive x_1-axis on the positive x_1-face (a surface with an outward normal pointing in the direction of the positive x_1-axis) of the surface element. The positiveness of other components on the same face are determined by the right-hand coordinate axis rule. For example, σ_{ij} are all positive if they are directed as shown in the diagram (figure 4.1). For a better understanding of the problem figure 4.1 should be studied carefully.

$\sigma_{11}, \sigma_{12}, \ldots$ are known as the components of stress. σ_{ii} are called normal stress components and σ_{ij} $(i \neq j)$ are called shear stress components. Each of these components has the dimension of force per unit area. We shall see presently that these behave as components of a tensor of rank two and so they are also called stress tensors. It is interesting to recall that tensor literally means stress. In fact, tensors and their transformation laws have their conceptual roots in the consideration of stress.

4.2 Expression for force due to stress

Consider a rectangular parallelepiped whose edges are parallel to the Cartesian axes x_1, x_2 and x_3. First imagine any two faces of this parallelepiped

† Though we use this convention throughout the book, the reader should remember that different conventions are sometimes used for these components by other authors.

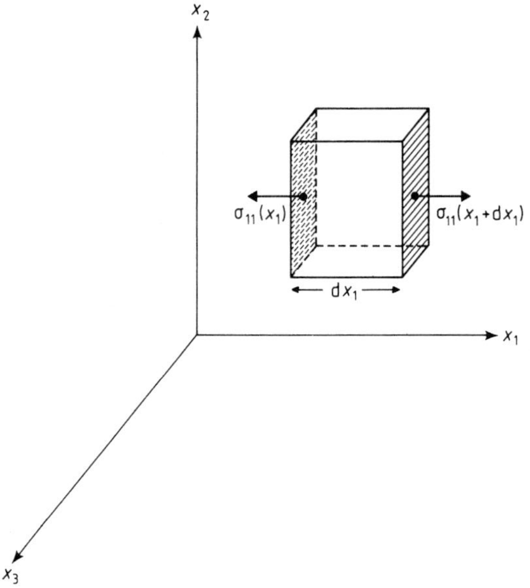

Figure 4.2 The stress on two faces of a parallelepiped along the x_1-direction.

along the x_1-axis say at x_1 and at an infinitesimal distance dx_1, i.e. at $x_1 + dx_1$. Assuming the continuity of stress let $\sigma_{11}(x_1)$ be the stress at x_1 and $\sigma_{11}(x_1 + dx_1)$ be the stress at $x_1 + dx_1$ and be directed as shown in figure 4.2. Therefore the x_1-component of the force acting on these faces would be $dx_2\, dx_3[\sigma_{11}(x_1 + dx_1) - \sigma_{11}(x_1)]$. In much the same way, the x_1-component of the force acting on the faces of the parallelepiped along the x_2- and the x_3-axes would be $dx_1\, dx_3[\sigma_{21}(x_2 + dx_2) - \sigma_{21}(x_2)]$ and $dx_1\, dx_2[\sigma_{31}(x_3 + dx_3) - \sigma_{31}(x_3)]$ respectively. Hence, the total x_1-component of the force acting on the parallelepiped is given by

$$dx_2\, dx_3[\sigma_{11}(x_1 + dx_1) - \sigma_{11}(x_1)] + dx_1\, dx_3[\sigma_{21}(x_2 + dx_2) - \sigma_{21}(x_2)] + dx_1\, dx_2[\sigma_{31}(x_3 + dx_3) - \sigma_{31}(x_3)]. \tag{4.1}$$

On making a Taylor expansion of $\sigma(x + dx)$ around x and retaining only the linear terms (higher-order terms are neglected as dx_1, dx_2 and dx_3 are infinitesimally small), the above equation becomes equal to

$$dx_1\, dx_2\, dx_3 \left(\frac{\partial \sigma_{11}}{\partial x_1} + \frac{\partial \sigma_{21}}{\partial x_2} + \frac{\partial \sigma_{31}}{\partial x_3}\right) \tag{4.2}$$

which is the expression for the x_1-component of the force acting on a volume element $dx_1 dx_2 dx_3$ of the parallelepiped due to the stress components σ_{ij}. Denoting the x_1-component of force per unit volume by $f_1^{(\sigma)}$, one has

$$f_1^{(\sigma)} = \frac{\partial \sigma_{11}}{\partial x_1} + \frac{\partial \sigma_{21}}{\partial x_2} + \frac{\partial \sigma_{31}}{\partial x_3}. \tag{4.3a}$$

Similar expressions for the x_2- and x_3-components of the force can also be obtained:

$$f_2^{(\sigma)} = \frac{\partial \sigma_{12}}{\partial x_1} + \frac{\partial \sigma_{22}}{\partial x_2} + \frac{\partial \sigma_{32}}{\partial x_3} \tag{4.3b}$$

$$f_3^{(\sigma)} = \frac{\partial \sigma_{13}}{\partial x_1} + \frac{\partial \sigma_{23}}{\partial x_2} + \frac{\partial \sigma_{33}}{\partial x_3}. \tag{4.3c}$$

In an abbreviated form, the *i*th component of the force per unit volume due to stress is

$$f_i^{(\sigma)} = \partial \sigma_{ji} / \partial x_j. \tag{4.4}$$

4.2.1 Proof that $\boldsymbol{\sigma}$ is a tensor of rank two

At this point it is simple to show that $\boldsymbol{\sigma}$ is a tensor of rank two. We recall from Chapter 1 that if T_{ij} are components of a tensor $\mathbf{T}$ of rank two then $\partial T_{ji}/\partial x_j$ are components of a vector,

$$B_i = \partial T_{ji} / \partial x_j. \tag{4.5}$$

Comparing equations (4.4) and (4.5) and remembering that the $f_i^{(\sigma)}$ are components of a vector by virtue of their physical significance, one infers that σ_{ij} are components of a tensor $\boldsymbol{\sigma}$ of rank two. A more direct approach showing that $\boldsymbol{\sigma}$ is a tensor of rank two is taken in Appendix **A.1**.

Since $\boldsymbol{\sigma}$ is a tensor of rank two, its components transform as (see equation (2.28))

$$\sigma'_{ij} = a_{im} a_{jn} \sigma_{mn}. \tag{4.6}$$

4.3 Conditions for static equilibrium

The state of static equilibrium of a material body requires that the resultant force and the moment are both zero. To facilitate further discussion the equilibrium conditions may be categorised as: (i) the body force condition, (ii) the moment condition, and (iii) the boundary condition at the surface.

4.3.1 Body force condition

Every material point in a body is always under the influence of body forces. These are gravitational forces which arise either due to the action of the other particles of the same or of another body. Let F_i denote the components of the body forces acting per unit mass and let ρ be the density (mass per unit volume) of the material, then the components of the body forces acting on the volume element $dx_1dx_2dx_3$ are $(\rho F_i)\, dx_1dx_2dx_3$. The body force condition requires that the total force, which is the sum of the body forces $(\rho F_i\, dx_1dx_2dx_3)$ and the stress forces $(f_i^{(\sigma)}\, dx_1dx_2dx_3)$, is zero,

$$(f_i^{(\sigma)} + \rho F_i)\, dx_1dx_2dx_3 = 0. \tag{4.7}$$

On substituting $f_i^{(\sigma)}$ from equation (4.4) one obtains the body force condition

$$\partial\sigma_{ji}/\partial x_j + \rho F_i = 0. \tag{4.8}$$

This is one of the most general forms of the equilibrium equation. We shall refer to it quite frequently in this book.

4.3.2 *Moment condition*

The moment condition requires that, in addition to (4.8), the moment of all the forces acting on the volume element along three perpendicular directions x_1, x_2, x_3 must be zero. First, we consider the moment of the stress forces, as shown in figure 4.3, acting on the faces of the parallelepiped parallel to the x_3-axis (positive sign is considered for anticlockwise rotation and negative sign is considered for clockwise rotation),

$$\begin{aligned}\tau^{(\sigma)} &= \sigma_{12}(dx_2dx_3)dx_1 - \sigma_{21}(dx_1dx_3)dx_2\\ &= (\sigma_{12} - \sigma_{21})dx_1dx_2dx_3.\end{aligned} \tag{4.9}$$

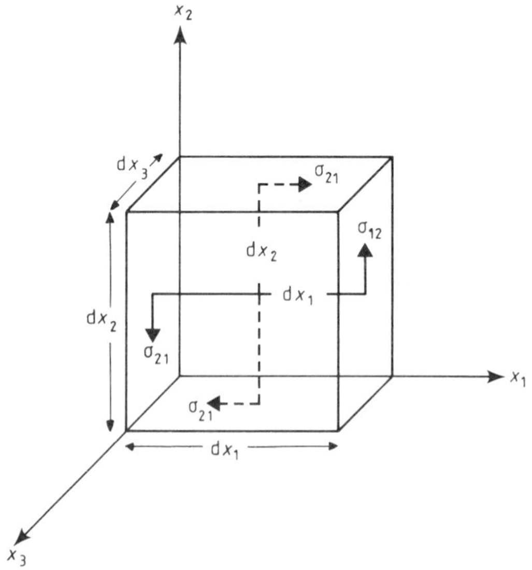

Figure 4.3 This diagram illustrates the moment of the stress forces acting at the parallelepiped faces parallel to the x_3-axis.

The moment due to body forces, $\boldsymbol{F}$, on the other hand, is necessarily of the order of $\tau^{(F)} \simeq (\rho \boldsymbol{F}\, dx_1 dx_2 dx_3) dl$, where dl represents an infinitesimal length. Since dl is very small and since ρ is also finite and small, $\tau^{(F)}$ (body moment) can be neglected in comparison with $\tau^{(\sigma)}$. Therefore, the vanishing moment condition becomes

$$(\sigma_{12}-\sigma_{21})\, dx_1 dx_2 dx_3 = 0. \tag{4.10}$$

Hence it follows that

$$\sigma_{12}=\sigma_{21}. \tag{4.11}$$

Similarly taking moments about the x_2- and x_1-axes one obtains

$$\sigma_{13}=\sigma_{31} \qquad \sigma_{23}=\sigma_{32}. \tag{4.12}$$

Equations (4.11) and (4.12) lead to a general result

$$\sigma_{ij}=\sigma_{ji}. \tag{4.13}$$

We conclude that the stress tensor is symmetric. This important result has been obtained assuming that the body moments do not exist. Strictly speaking, equation (4.13) holds true for a solid as long as the interatomic forces are such that no internal body moments exist. The absence of such body moments is a fundamental assumption in the theory of elasticity. This has been a point of discussion from time to time. However, it may be noted that there are no sufficient reasons to assume that the interatomic forces are such that the body moment is zero. As a matter of fact a theory can be developed formally without assuming $\sigma_{ij}=\sigma_{ji}$ which is definitely a topic beyond the scope of the present book.

4.3.3 Boundary condition at the surface

The stress or tension in a body is created by applying external forces at the surface. For example, in the Young's modulus experiment the tension is created by putting a load at the end of the wire. The external forces acting at the surface are usually called surface forces. Presently we discuss the equilibrium at every point of the surface in terms of surface forces.

Consider a surface element of area dA of a material body which is subjected to a surface force $\mathscr{P}$ acting per unit area. This creates stress and let $\mathscr{F}^{(\sigma)}$ denote the force developed at the surface per unit area due to stress σ. For equilibrium at every point of the surface,

$$\mathscr{P}_i\, dA - \mathscr{F}_i^{(\sigma)}\, dA = 0 \tag{4.14}$$

or

$$\mathscr{F}_i^{(\sigma)} = \mathscr{P}_i \qquad i=1, 2, 3. \tag{4.15}$$

Below we give a simple derivation of $\mathscr{F}_i^{(\sigma)}$ in terms of σ_{ij} employing the transformation equation (2.1*b*). Suppose σ'_{11}, σ'_{12} and σ'_{13} are the stress components in the primed frame of reference such that $\sigma'_{11} \| x'_1$, $\sigma'_{12} \| x'_2$ and $\sigma'_{13} \| x'_3$. If $\mathscr{F}_1^{(\sigma)}$, $\mathscr{F}_2^{(\sigma)}$ and $\mathscr{F}_3^{(\sigma)}$ represent the force components per unit area acting along the x_1-, x_2- and x_3-axes in the unprimed frame, then

$$\mathscr{F}_1^{(\sigma)} = \sigma'_{11} a_{11} + \sigma'_{12} a_{21} + \sigma'_{13} a_{31} \tag{4.16a}$$

$$\mathscr{F}_2^{(\sigma)} = \sigma'_{11} a_{12} + \sigma'_{12} a_{22} + \sigma'_{13} a_{32} \tag{4.16b}$$

$$\mathscr{F}_3^{(\sigma)} = \sigma'_{11} a_{13} + \sigma'_{12} a_{23} + \sigma'_{13} a_{33} \tag{4.16c}$$

or in short,

$$\mathscr{F}_i^{(\sigma)} = \sigma'_{11} a_{1i} + \sigma'_{12} a_{2i} + \sigma'_{13} a_{3i}. \tag{4.17}$$

Substituting for σ'_{ij} from equation (4.6), the above equation becomes

$$\mathscr{F}_i^{(\sigma)} = \sum_{mn} (a_{1m} a_{1n} \sigma_{mn} a_{1i} + a_{1m} a_{2n} \sigma_{mn} a_{2i} + a_{1m} a_{3n} \sigma_{mn} a_{3i})$$

$$= \sum_{mn} \sigma_{mn} a_{1m} (a_{1n} a_{1i} + a_{2n} a_{2i} + a_{3n} a_{3i})$$

$$= \sum_{mn} \sigma_{mn} a_{1m} \delta_{ni}$$

$$= \sum_{m} \sigma_{mi} a_{1m} \qquad (\because \quad \delta_{ni} = 1 \text{ for } n = i).$$

Therefore,

$$\mathscr{F}_i^{(\sigma)} = \sum_{m} a_{1m} \sigma_{mi}$$

$$= a_{11} \sigma_{1i} + a_{12} \sigma_{2i} + a_{13} \sigma_{3i}. \tag{4.18}$$

If we now denote $\boldsymbol{n}$ as a unit vector normal to the surface area $\mathrm{d}A$ and parallel to x'_1, then identifying a_{11} as n_1, a_{12} as n_2 and a_{13} as n_3 the equation (4.18) reduces to

$$\mathscr{F}_i^{(\sigma)} = n_1 \sigma_{1i} + n_2 \sigma_{2i} + n_3 \sigma_{3i} \tag{4.19}$$

or,

$$\mathscr{F}_i^{(\sigma)} = n_j \sigma_{ji} \tag{4.20}$$

the desired relation between the stress vector $\mathscr{F}_i^{(\sigma)}$ and the stress components σ_{ij}. Equation (4.20) is also called Cauchy's formula. Substituting this in equation (4.15), the equilibrium condition at the surface becomes

$$n_j \sigma_{ji} = \mathscr{P}_i. \tag{4.21}$$

Note that if no surface force ($\mathscr{P}_i = 0$) is acting at the surface, then this is a free surface, so that the equilibrium condition is simply

$$n_j\sigma_{ji}=0. \tag{4.22}$$

Also, if one considers the equilibrium of two media say (a) and (b) in contact then across the boundary one must have

$$n_j\sigma_{ji}^{(a)}=n_j\sigma_{ji}^{(b)} \tag{4.23}$$

where $\sigma_{ji}^{(a)}$ and $\sigma_{ji}^{(b)}$ are the stress components at the surface layer of the boundary in media (a) and (b) respectively.

4.4 Expression for the work done on a solid body by stresses and body forces

A solid body subjected to external forces or stresses undergoes deformations. The stress tensor is intimately related to the deformation tensor, but in what way? Before we answer this question, it is instructive to calculate the amount of work that has to be done on a solid to bring about a deformation such that a displacement $\boldsymbol{s}(\boldsymbol{x})$ at a point $\boldsymbol{x}$ changes by a small amount $\Delta\boldsymbol{s}(\boldsymbol{x})$.

Consider an element of volume $\mathrm{d}x_1\mathrm{d}x_2\mathrm{d}x_3$ of a solid body in the form of a parallelepiped (see figure 4.3). As before, we assume that it is acted upon by the stress components σ_{ij} and a body force $\boldsymbol{F}$. These produce deformations at each point which have components along the x_1-, x_2- and x_3-axes. First, we consider the x_1-component of the change in displacement, $\Delta s_1(\boldsymbol{x})$, and the associated amount of work done on the element,

$$\begin{aligned}\mathrm{d}w_1\mathrm{d}x_1\mathrm{d}x_2\mathrm{d}x_3=(\rho F_1\mathrm{d}x_1\mathrm{d}x_2\mathrm{d}x_3)\,\Delta s_1+\sigma_{11}(x_1+\mathrm{d}x_1)\,\mathrm{d}x_2\mathrm{d}x_3\\ \times\Delta s_1(x_1+\mathrm{d}x_1)-\sigma_{11}(x_1)\,\mathrm{d}x_2\,\mathrm{d}x_3\,\Delta s_1(x_1)\\ +\sigma_{21}(x_2+\mathrm{d}x_2)\,\mathrm{d}x_1\,\mathrm{d}x_3\,\Delta s_1(x_2+\mathrm{d}x_2)\\ -\sigma_{21}(x_2)\,\mathrm{d}x_1\mathrm{d}x_3\,\Delta s_1(x_2)+\sigma_{31}(x_3+\mathrm{d}x_3)\,\mathrm{d}x_1\mathrm{d}x_2\\ \times\Delta s_1(x_3+\mathrm{d}x_3)-\sigma_{31}(x_3)\,\mathrm{d}x_1\,\mathrm{d}x_2\,\Delta s_1(x_3).\end{aligned} \tag{4.24}$$

The first term is due to the body force and the others are due to stress components. $\sigma(\boldsymbol{x})$ and $\Delta s_1(\boldsymbol{x})$ represent the stress and the change in displacement at the face $\boldsymbol{x}$. Rearranging the terms in equation (4.24),

$$\begin{aligned}\mathrm{d}w_1\mathrm{d}x_1\mathrm{d}x_2\mathrm{d}x_3=(\rho F_1\,\Delta s_1)\,\mathrm{d}x_1\mathrm{d}x_2\mathrm{d}x_3\\ +[(\sigma_{11}\,\Delta s_1)_{x_1+\mathrm{d}x_1}-(\sigma_{11}\,\Delta s_1)_{x_1}]\,\mathrm{d}x_2\,\mathrm{d}x_3\\ +[(\sigma_{21}\,\Delta s_1)_{x_2+\mathrm{d}x_2}-(\sigma_{21}\,\Delta s_1)_{x_2}]\,\mathrm{d}x_1\,\mathrm{d}x_3\\ +[(\sigma_{31}\,\Delta s_1)_{x_3+\mathrm{d}x_3}-(\sigma_{31}\,\Delta s_1)_{x_3}]\,\mathrm{d}x_1\,\mathrm{d}x_2.\end{aligned} \tag{4.25}$$

After a Taylor expansion up to linear terms in Δs_1, one has

$$dw_1 dx_1 dx_2 dx_3 = (\rho F_1 \, \Delta s_1) \, dx_1 dx_2 dx_3 + \left(\frac{\partial(\sigma_{11} \, \Delta s_1)}{\partial x_1} + \frac{\partial(\sigma_{21} \, \Delta s_1)}{\partial x_2} + \frac{\partial(\sigma_{31} \, \Delta s_1)}{\partial x_3} \right) dx_1 dx_2 dx_3$$

which is equivalent to

$$dw_1 = \rho F_1 \, \Delta s_1 + \frac{\partial(\sigma_{11} \, \Delta s_1)}{\partial x_1} + \frac{\partial(\sigma_{21} \, \Delta s_1)}{\partial x_2} + \frac{\partial(\sigma_{31} \, \Delta s_1)}{\partial x_3}. \tag{4.26}$$

In an abbreviated form, one writes

$$dw_1 = \rho F_1 \, \Delta s_1 + \partial(\sigma_{j1} \, \Delta s_1)/\partial x_j. \tag{4.27}$$

Rewriting the second term on the right-hand side as

$$\frac{\partial(\sigma_{j1} \, \Delta s_1)}{\partial x_j} = \Delta s_1 \left(\frac{\partial \sigma_{j1}}{\partial x_j} \right) + \sigma_{j1} \left(\frac{\partial \, \Delta s_1}{\partial x_j} \right) \tag{4.28}$$

equation (4.27) becomes,

$$dw_1 = \left(\rho F_1 + \frac{\partial \sigma_{j1}}{\partial x_j} \right) \Delta s_1 + \sigma_{j1} \frac{\partial \, \Delta s_1}{\partial x_j}. \tag{4.29}$$

Similar expressions for dw_2 and dw_3 associated with small displacements Δs_2 and Δs_3 along the x_2- and x_3-axes can also be obtained:

$$dw_2 = \left(\rho F_2 + \frac{\partial \sigma_{j2}}{\partial x_j} \right) \Delta s_2 + \sigma_{j2} \frac{\partial \, \Delta s_2}{\partial x_j} \tag{4.30}$$

$$dw_3 = \left(\rho F_3 + \frac{\partial \sigma_{j3}}{\partial x_j} \right) \Delta s_3 + \sigma_{j3} \frac{\partial \, \Delta s_3}{\partial x_j}. \tag{4.31}$$

Adding expressions (4.29) to (4.31), one obtains an expression for dw, the total amount of work done per unit volume,

$$dw = dw_1 + dw_2 + dw_3 = \sum_i \left(\rho F_i + \frac{\partial \sigma_{ji}}{\partial x_j} \right) \Delta s_i + \sum_{ji} \sigma_{ji} \frac{\partial \, \Delta s_i}{\partial x_j}. \tag{4.32}$$

If the solid under consideration is in a state of static equilibrium, then the terms in brackets in equation (4.32) contribute nothing (see body force condition (4.8)). Hence, for a quasi-static process, equation (4.32) becomes

$$dw = \sigma_{ji} \, \partial \, \Delta s_i / \partial x_j \tag{4.33}$$

which is just the work done by the stresses. In view of the moment condition, $\sigma_{ij} = \sigma_{ji}$, the above equation can also be written as

$$dw = \sigma_{ji} \left[\frac{1}{2} \left(\frac{\partial \, \Delta s_i}{\partial x_j} + \frac{\partial \, \Delta s_j}{\partial x_i} \right) \right]. \tag{4.34}$$

Now if we write

$$\partial\, \Delta s_i/\partial x_j = \Delta\, \partial s_i/\partial x_j \tag{4.35}$$

then using equations (4.34) and (3.12) one has

$$\mathrm{d}w = \sigma_{ji}\, \Delta\varepsilon_{ij} = \sigma_{ij}\, \Delta\varepsilon_{ij}. \tag{4.36}$$

The summations over i and j are implied. Thus, the work done which is associated with the deformation of the system in a quasi-static process (the system is in equilibrium at every instant of time) depends only on the changes in the strain tensors ε_{ij} provided that σ_{ij} are symmetrical. Note that for $\sigma_{ij} \neq \sigma_{ji}$, the expression for d$w$ will contain not only the symmetric strain tensors ε_{ij} but also the antisymmetric strain tensors t_{ij}.

4.5 Strain-energy function and Voigt notation for stress and strain components

The principle of conservation of energy dictates that the total work done on a system by external forces is equal to the increase in its energy less the amount of heat absorbed by the system. For the moment if we ignore the heat absorbed, then the work done is simply utilised to increase its energy in the form of kinetic and internal energies. If the process is so slow that the kinetic energy can also be ignored, then the work done is equal to the change in internal energy. The latter is stored up in the system as strain energy. An equivalent mathematical expressions is given in equation (4.36). The change in energy, dw, is a function of the changes in the strain tensors ε_{ij}. w is called the strain-energy function and dw is a perfect differential;

$$\partial w/\partial\varepsilon_{ij} = \sigma_{ij} \tag{4.37}$$

and hence

$$\frac{\partial^2 w}{\partial\varepsilon_{ij}\,\partial\varepsilon_{kl}} = \frac{\partial^2 w}{\partial\varepsilon_{kl}\,\partial\varepsilon_{ij}} = \frac{\partial\sigma_{ij}}{\partial\varepsilon_{kl}} = \frac{\partial\sigma_{kl}}{\partial\varepsilon_{ij}}. \tag{4.38}$$

At this point it is convenient to introduce a new notation (Voigt 1910) for stress and strain components, as

$$\begin{array}{llll}
\sigma_{11}=\sigma_1 & \sigma_{23}\ (=\sigma_{32})=\sigma_4 & \text{and}\ \ \varepsilon_{11}=\varepsilon_1 & 2\varepsilon_{23}\ (=2\varepsilon_{32})=\varepsilon_4 \\
\sigma_{22}=\sigma_2 & \sigma_{31}\ (=\sigma_{13})=\sigma_5 & \varepsilon_{22}=\varepsilon_2 & 2\varepsilon_{13}\ (=2\varepsilon_{31})=\varepsilon_5 \\
\sigma_{33}=\sigma_3 & \sigma_{12}\ (=\sigma_{21})=\sigma_6 & \varepsilon_{33}=\varepsilon_3 & 2\varepsilon_{12}\ (=2\varepsilon_{21})=\varepsilon_6.
\end{array} \tag{4.39}$$

Thus, after taking account of the symmetry of the stress and strain components we have, in general, six stress components and six strain components.

Employing this Voigt notation, equation (4.36) becomes

$$dw = \sigma_1\, d\varepsilon_1 + \sigma_2\, d\varepsilon_2 + \sigma_3\, d\varepsilon_3 + \sigma_4\, d\varepsilon_4 + \sigma_5\, d\varepsilon_5 + \sigma_6\, d\varepsilon_6 \tag{4.40}$$

or in an abbreviated form:

$$dw = \sum_{m=1}^{6} \sigma_m\, d\varepsilon_m. \tag{4.41}$$

Therefore, the conditions that dw is a perfect differential are

$$\begin{aligned} &\partial w/\partial \varepsilon_m = \sigma_m \\ &\frac{\partial^2 w}{\partial \varepsilon_m\, \partial \varepsilon_n} = \frac{\partial^2 w}{\partial \varepsilon_n\, \partial \varepsilon_m} = \frac{\partial \sigma_m}{\partial \varepsilon_n} = \frac{\partial \sigma_n}{\partial \varepsilon_m} \end{aligned} \qquad m = n = 1, 6. \tag{4.42}$$

4.6 Generalised Hooke's law and elastic constants

For most solids it is experimentally found that the measured extension (strain) is proportional to the applied load (stress) provided the load does not exceed a value known as the elastic limit; this is Hooke's law. Generalising this, one states that in the elastic range, each of the six stress components at a given point is a linear function of the six strain components at that point. Using the Voigt notation, one writes

$$\sigma_m = \sum_{n=1}^{6} c_{mn}\varepsilon_n \qquad m = 1, 6 \tag{4.43}$$

or conversely

$$\varepsilon_m = \sum_{n=1}^{6} s_{mn}\sigma_n \qquad m = 1, 6 \tag{4.44}$$

where c_{mn} are called elastic stiffness constants (or simply the elastic constants) and s_{mn} are called elastic compliance constants. Each of equations (4.43) and (4.44) represents six equations each with six added terms on the right-hand side. Both c_{mn} and s_{mn} form 6×6 matrices and each of them has, in general, 36 independent numbers. Explicitly,

$$c_{mn} = \begin{pmatrix} c_{11} & c_{12} & c_{13} & c_{14} & c_{15} & c_{16} \\ c_{21} & c_{22} & c_{23} & c_{24} & c_{25} & c_{26} \\ c_{31} & c_{32} & c_{33} & c_{34} & c_{35} & c_{36} \\ c_{41} & c_{42} & c_{43} & c_{44} & c_{45} & c_{46} \\ c_{51} & c_{52} & c_{53} & c_{54} & c_{55} & c_{56} \\ c_{61} & c_{62} & c_{63} & c_{64} & c_{65} & c_{66} \end{pmatrix} \tag{4.45}$$

with a similar result for s_{mn}. From equations (4.45) and (4.44),

$$\varepsilon_m = s_{mn}\sigma_n = s_{mn}c_{nl}\varepsilon_l \tag{4.46}$$

which shows that s_{mn} and c_{nl} satisfy the orthogonality relations

$$s_{mn}c_{nl} = \delta_{ml}. \tag{4.47}$$

Thus s_{mn} are uniquely determined if c_{nl} are known and *vice versa*. In addition, because of equation (4.42), i.e.

$$\partial\sigma_m/\partial\varepsilon_n = \partial\sigma_n/\partial\varepsilon_m$$

it follows from equations (4.43) and (4.44) that

$$c_{mn} = c_{nm} \qquad s_{mn} = s_{nm}. \tag{4.48}$$

Hence both c_{mn} and s_{mn} are symmetrical and, in effect, the number of independent elements is reduced from 36 to 21. For different crystal symmetries the numbers are reduced further depending upon the symmetry of the crystal. We shall take up this problem in some detail in Chapter 7. The maximum reduction is possible for elastically isotropic materials. These materials are very useful from the engineering and geophysical points of view and we therefore prefer to discuss them separately in the following Part II of the Chapter.

PART II. ELASTIC CONSTANTS FOR ISOTROPIC SOLIDS AND FOR VARIOUS HOMOGENEOUS DEFORMATIONS

4.7 Elastic constants for isotropic solids

4.7.1 Strain-energy function for isotropic solids

A solid body is said to be elastically isotropic if its elastic properties are the same in all directions. There are only two independent elastic constants for isotropic materials. We shall show this from considerations on the strain-energy function w. On substituting σ_m from equation (4.43) into equation (4.41) one has

$$\mathrm{d}w = \sum_{m,n=1}^{6} (c_{mn}\varepsilon_n)\,\mathrm{d}\varepsilon_m. \tag{4.49}$$

It follows immediately that

$$\begin{aligned} w &= \frac{1}{2}\sum_{m,n=1}^{6} c_{mn}\varepsilon_m\varepsilon_n \\ &= \tfrac{1}{2}(c_{11}\varepsilon_1^2 + c_{22}\varepsilon_2^2 + \cdots + 2c_{12}\varepsilon_1\varepsilon_2 + \cdots). \end{aligned} \tag{4.50}$$

Therefore,

$$\frac{\partial w}{\partial \varepsilon_m} = \sum_{n=1}^{6} c_{mn}\varepsilon_n \qquad \text{and} \qquad \frac{\partial^2 w}{\partial \varepsilon_m\, \partial \varepsilon_n} = c_{mn}. \tag{4.51}$$

Thus, w is a quadratic function of the strain components ε_m. Let us now recall what one means by isotropy. Mathematically speaking it implies no orientation effect in the material. Thus, w must be invariant with respect to a rotation of the coordinates. In other words, w is a function of those combinations of ε_m which are invariant. Bearing in mind equation (2.74), the invariant conditions associated with ε_m are:

$$\Delta = \Delta' = \varepsilon_1 + \varepsilon_2 + \varepsilon_3 \tag{4.52}$$

$$\Theta = \Theta' = \varepsilon_1\varepsilon_2 + \varepsilon_2\varepsilon_3 + \varepsilon_3\varepsilon_1 - \tfrac{1}{4}(\varepsilon_4^2 + \varepsilon_5^2 + \varepsilon_6^2). \tag{4.53}$$

We, therefore, express w in mathematical form as

$$w = a\,\Delta^2 + b\Theta. \tag{4.54}$$

The constants a and b are characteristics of the elastic isotropy.

4.7.2 Lame's elastic constants

Equation (4.54) is frequently written as

$$w = \tfrac{1}{2}\lambda\,\Delta^2 + \mu(\Delta^2 - 2\Theta) \tag{4.55}$$

where λ and μ are called the Lame elastic constants. If we express Δ and Θ in tensor notation,

$$\Delta = \sum_{i=1}^{3} \varepsilon_{ii} \tag{4.56}$$

$$\Delta^2 - 2\Theta = \sum_{i,j=1}^{3} \varepsilon_{ij}^2 \tag{4.57}$$

then (4.55) becomes

$$w = \tfrac{1}{2}\lambda\,\Delta^2 + \mu \sum_{i,j}^{3} \varepsilon_{ij}^2. \tag{4.58}$$

Using this and equation (4.37), one obtains various stress components,

$$\sigma_{11} = \lambda\,\Delta + 2\mu\varepsilon_{11} \tag{4.59}$$

$$\sigma_{22} = \lambda\,\Delta + 2\mu\varepsilon_{22} \tag{4.60}$$

$$\sigma_{33} = \lambda\,\Delta + 2\mu\varepsilon_{33} \tag{4.61}$$

$$\sigma_{23} = 2\mu\varepsilon_{23} \tag{4.62}$$

$$\sigma_{13} = 2\mu\varepsilon_{13} \tag{4.63}$$

$$\sigma_{12} = 2\mu\varepsilon_{12}. \tag{4.64}$$

In an abbreviated form equations (4.59) to (4.64) can be written as

$$\sigma_{ij} = 2\mu\varepsilon_{ij} + \lambda\,\Delta\delta_{ij}. \tag{4.65}$$

This in conjunction with Hooke's law yields elastic constants for isotropic solids,

$$c_{mn} = \begin{pmatrix} \lambda+2\mu & \lambda & \lambda & 0 & 0 & 0 \\ \lambda & \lambda+2\mu & \lambda & 0 & 0 & 0 \\ \lambda & \lambda & \lambda+2\mu & 0 & 0 & 0 \\ 0 & 0 & 0 & \mu & 0 & 0 \\ 0 & 0 & 0 & 0 & \mu & 0 \\ 0 & 0 & 0 & 0 & 0 & \mu \end{pmatrix} \tag{4.66}$$

Therefore, the non-vanishing elastic constants for isotropic solids are

$$c_{11} = c_{22} = c_{33} = \lambda + 2\mu \tag{4.67}$$

$$c_{12} = c_{13} = c_{23} = \lambda \tag{4.68}$$

$$c_{44} = c_{55} = c_{66} = \mu. \tag{4.69}$$

From these one infers that

$$2c_{44} = c_{11} - c_{12}. \tag{4.70}$$

Thus we are left with only two independent elastic constants, namely c_{12} and c_{44} for isotropic materials. These are respectively Lame's constants λ and μ. We shall see later in Chapter 7 that c_{11}, c_{12} and c_{44} are also three independent elastic constants for cubic materials. The condition (4.70) is, however, no longer true. The quantity $2c_{44}/(c_{11}-c_{12})$ is often used as a measure of anisotropy.

At this stage it is advantageous to write down the elastic compliance constants s_{mn} in terms of λ and μ. This can easily be accomplished bearing in mind the orthogonality relation (4.47). Writing this explicitly,

$$m=1,\ l=1 \qquad c_{11}s_{11} + c_{12}s_{21} + c_{13}s_{31} = 1 \tag{4.71}$$

$$m=1,\ l=2 \qquad c_{11}s_{12} + c_{12}s_{22} + c_{13}s_{32} = 0 \tag{4.72}$$

$$c_{44}s_{44} = 1. \tag{4.73}$$

These, when used with the isotropic conditions $s_{11} = s_{22} = s_{33}$; $s_{12} = s_{13} = s_{23}$; $s_{44} = s_{55} = s_{66}$, yield

$$s_{11} = \frac{c_{11} + c_{12}}{(c_{11} - c_{12})(c_{11} + 2c_{12})} = \frac{\lambda + \mu}{\mu(3\lambda + 2\mu)} \tag{4.74}$$

$$s_{12} = \frac{-c_{12}}{(c_{11} - c_{12})(c_{11} + 2c_{12})} = -\frac{\lambda}{2\mu(3\lambda + 2\mu)} \tag{4.75}$$

$$s_{44} = 1/c_{44} = \mu^{-1}. \tag{4.76}$$

Thus one also has

$$s_{44}=2(s_{11}-s_{12}) \tag{4.77}$$

which is a counterpart of equation (4.70).

We conclude this section by writing c_{11}, c_{12} and c_{44} in terms of s_{11}, s_{12} and s_{44},

$$c_{11}=\frac{s_{11}+s_{12}}{(s_{11}-s_{12})(s_{11}+2s_{12})}=\lambda+2\mu \tag{4.78}$$

$$c_{12}=-\frac{s_{12}}{(s_{11}-s_{12})(s_{11}+2s_{12})}=\lambda \tag{4.79}$$

$$c_{44}=1/s_{44}=\mu. \tag{4.80}$$

4.8 Some simple cases of homogeneous deformations and elastic moduli

4.8.1 Young's modulus and Poisson's ratio

When a wire of linear dimensions is subjected to uniaxial stress, say along the x_1-axis, then all the $\sigma_n=0$ except for σ_1. For cubic and isotropic materials, the strain–stress relation (4.44) becomes

$$\varepsilon_1=s_{11}\sigma_1 \tag{4.81}$$

$$\varepsilon_2=s_{21}\sigma_1=s_{12}\sigma_1 \tag{4.82}$$

$$\varepsilon_3=s_{31}\sigma_1=s_{13}\sigma_1 \tag{4.83}$$

$$\varepsilon_4=\varepsilon_5=\varepsilon_6=0. \tag{4.84}$$

Thus we see that even if the stress is applied unidirectionally, the deformation occurs in all possible directions. The ratio of the longitudinal stress σ_1 and the longitudinal strain ε_1 measured in the same direction is called Young's modulus of elasticity, E.

$$E=\frac{\sigma_1}{\varepsilon_1}=\frac{1}{s_{11}}. \tag{4.85}$$

For isotropic materials E can be written in terms of λ and μ using the relation (4.74),

$$E=\mu(3\lambda+2\mu)/(\lambda+\mu). \tag{4.86}$$

It should be noted that in the Young's modulus experiment σ_1 produces not only the longitudinal extension ε_1 but also the lateral contractions ε_2 and ε_3. If we use a negative sign to represent contractions, equations (4.82) and (4.83) can also be written as

$$-\varepsilon_2 = s_{12}\sigma_1 \qquad -\varepsilon_3 = s_{13}\sigma_1. \tag{4.87}$$

For isotropic and cubic crystals $s_{12}=s_{13}$ and hence the lateral contraction is equal to $-\varepsilon_2=-\varepsilon_3$. The ratio of the lateral contraction to the corresponding longitudinal extension is called Poisson's ratio,

$$\nu = -\frac{\varepsilon_2}{\varepsilon_1} = -\frac{\varepsilon_3}{\varepsilon_1} = -\frac{s_{12}}{s_{11}}. \tag{4.88}$$

Substituting for s_{12} and s_{11} from equations (4.75) and (4.74), one has

$$\nu = \frac{c_{12}}{c_{11}+c_{12}} = \frac{\lambda}{2(\lambda+\mu)}. \tag{4.89}$$

Poisson in 1829 argued for a universal value of $\nu=\frac{1}{4}$ for all materials. This makes $\lambda=\mu$ and hence $c_{12}=c_{44}$. In such a case the elastic behaviour of an isotropic solid is described by just one independent elastic constant. It must be noted, however, that Poisson's ratio ν deviates from a quarter for most solids.

Furthermore, equations (4.86) and (4.89) can be used to obtain Lame's constants in terms of E and ν as

$$\lambda = \frac{E\nu}{(1-2\nu)(1+\nu)} \qquad \text{and } \mu = \frac{E}{2(1+\nu)}. \tag{4.90}$$

4.8.2 Bulk modulus and compressibility

These quantities relate the uniform hydrostatic compression to the corresponding fractional change in volume. In particular, the bulk modulus, B, is defined as

$$B = -\frac{p}{\Delta V/V} \tag{4.91}$$

where ΔV is the change in volume for a given volume V when uniform pressure p is applied. For uniform hydrostatic compression, the pressure p in terms of the stress σ_{ij} is expressed as

$$\sigma_{ij} = -p\delta_{ij} \tag{4.92}$$

and therefore all σ_{ij} are zero except (in the Voigt notion)

$$\sigma_1 = \sigma_2 = \sigma_3 = -p. \tag{4.93}$$

Thus one defines for a solid, a mean pressure p as

$$p = -\tfrac{1}{3}(\sigma_1+\sigma_2+\sigma_3) = -\tfrac{1}{3}\sigma_i. \tag{4.94}$$

Also for solids having cubic symmetry $(\varepsilon_4=\varepsilon_5=\varepsilon_6=0)$ one has (see equation (4.43))

$$\begin{aligned}\sigma_1&=c_{11}\varepsilon_1+c_{12}\varepsilon_2+c_{13}\varepsilon_3+0+0+0\\ \sigma_2&=c_{21}\varepsilon_1+c_{22}\varepsilon_2+c_{23}\varepsilon_3+0+0+0\\ \sigma_3&=c_{31}\varepsilon_1+c_{32}\varepsilon_2+c_{33}\varepsilon_3+0+0+0\end{aligned} \tag{4.95}$$

which on using the cubic symmetry conditions, $c_{12}=c_{13}=c_{23}$ and $c_{11}=c_{22}=c_{33}$, yield

$$\sigma_1+\sigma_2+\sigma_3=(c_{11}+2c_{12})(\varepsilon_1+\varepsilon_2+\varepsilon_3). \tag{4.96}$$

The second factor on the right-hand side is simply the dilatation $\Delta\ (\equiv\Delta V/V)$ and hence (4.96) becomes

$$\sigma_1+\sigma_2+\sigma_3=(c_{11}+2c_{12})\ \Delta V/V. \tag{4.97}$$

The bulk modulus B is then

$$B=\tfrac{1}{3}(c_{11}+2c_{12}). \tag{4.98}$$

or in terms of Lame's constants one obtains

$$B=\lambda+\tfrac{2}{3}\mu=E/[3(1-2\nu)]. \tag{4.99}$$

The compressibility, K, is defined as the inverse of the bulk modulus B,

$$K=1/B=-\Delta V/pV. \tag{4.100}$$

The expression for K in terms of the elastic constants can be obtained straightforwardly by taking the inverse of equation (4.98), however, we prefer to proceed as follows. The strain–stress relation (4.44) for isotropic and cubic crystals is explicitly

$$\varepsilon_1=s_{11}\sigma_1+s_{12}\sigma_2+s_{13}\sigma_3 \tag{4.101}$$

$$\varepsilon_2=s_{21}\sigma_1+s_{22}\sigma_2+s_{23}\sigma_3 \tag{4.102}$$

$$\varepsilon_3=s_{31}\sigma_1+s_{32}\sigma_2+s_{33}\sigma_3. \tag{4.103}$$

Adding the above equations and using the symmetry conditions ($s_{11}=s_{22}=s_{33}$; $s_{12}=s_{21}=s_{31}=s_{13}=s_{23}=s_{32}$), one obtains

$$\Delta=\Delta V/V=\varepsilon_1+\varepsilon_2+\varepsilon_3=(s_{11}+2s_{12})(-3p)$$

or

$$-\Delta V/pV=3(s_{11}+2s_{12}). \tag{4.104}$$

Therefore, the compressibility, K, becomes

$$K=3(s_{11}+2s_{12})$$

or in terms of Lame's constants

$$K=\frac{3}{3\lambda+2\mu}=\frac{3(1-2\nu)}{E}. \tag{4.105}$$

4.8.3 Shear modulus

The shear modulus of elasticity defines the characteristic property of the material when the body is subjected to a pure-shear deformation. The ratio of the shear stress and the total angular change (see figure 3.4) produced by the stress is called the shear modulus,

$$G=\sigma_{12}/2\gamma=\sigma_{12}/2\varepsilon_{12}. \tag{4.106}$$

On using equation (4.64) it follows that

$$G=\mu. \tag{4.107}$$

Hence Lame's constant, μ, is just the shear modulus of elasticity which is also known as the modulus of rigidity or the torsion modulus.

We conclude this section by providing numerical values of the elastic moduli for some common materials in table 4.1. It is evident that the values of different elastic moduli differ appreciably from material to material. Such differences in the elastic properties of the materials make them interesting for diverse technical applications.

Table 4.1 Numerical values† of elastic moduli at room temperature for several materials. Each of E, G or B is expressed in units of $10^{11}\,\mathrm{dyn\,cm^{-2}}$.

	E	G	B	ν
Potassium	0.45	0.17	0.36	0.29
Ice (257 K)	0.94	0.36	0.89	0.32
Lead	2.81	1.01	4.27	0.39
Sodium chloride	3.60	1.44	2.39	0.25
Aluminum	7.08	2.63	7.75	0.35
Quartz	9.97	4.67	38.40	0.06
Copper	12.77	4.76	13.33	0.34
Iron	21.86	8.54	16.50	0.28
Beryllium	30.77	13.77	13.39	0.11
Diamond	102.84	47.22	41.70	0.09

† Data are taken from the compilation by Simmons and Wang (1971).

4.9 Crystal stability

For a crystal to be elastically stable the strain-energy function, w, has to be positive-definite, i.e. for any arbitrary strains, $w>0$. Hence the determinant of c_{mn} and all its co-minors must be greater than zero. This puts restrictions on the values of the various elastic moduli.

For an isotropic body we recall the strain-energy function (see equation (4.58))

$$w = \tfrac{1}{2}\lambda\, \Delta^2 + \mu \sum_{i,j}^{3} \varepsilon_{ij}^2$$

which is invariant under orthogonal transformations and hence can be referred to any choice of coordinate axes. Referring to the principal axes, w becomes

$$w = \tfrac{1}{2}\lambda(\varepsilon^{(1)} + \varepsilon^{(2)} + \varepsilon^{(3)})^2 + \mu[(\varepsilon^{(1)})^2 + (\varepsilon^{(2)})^2 + (\varepsilon^{(3)})^2].$$

It can also be written as

$$w = \tfrac{1}{2}(\lambda + \tfrac{2}{3}\mu)(\varepsilon^{(1)} + \varepsilon^{(2)} + \varepsilon^{(3)})^2 + \tfrac{1}{3}\mu[(\varepsilon^{(1)} - \varepsilon^{(2)})^2 + (\varepsilon^{(2)} - \varepsilon^{(3)})^2 + (\varepsilon^{(3)} - \varepsilon^{(1)})^2]. \quad (4.108)$$

From thermodynamic considerations w must be positive-definite irrespective of the nature of the deformations. Restricting ourselves, first of all, to shear deformations, i.e. $\varepsilon^{(1)} + \varepsilon^{(2)} + \varepsilon^{(3)} = 0$, equation (4.108) implies that

$$\mu > 0. \quad (4.109)$$

If the deformations, on the other hand, are such that $\varepsilon^{(1)} = \varepsilon^{(2)} = \varepsilon^{(3)}$, then one must have

$$\lambda + \tfrac{2}{3}\mu > 0 \qquad \text{or } B > 0. \quad (4.110)$$

Making use of the expressions (4.99) and (4.90) for B and μ, one writes for the stability condition:

$$\frac{E}{3(1-2\nu)} > 0 \qquad \text{and} \qquad \frac{E}{2(1+\nu)} > 0 \quad (4.111)$$

or equivalently,

$$E > 0 \qquad -1 < \nu < \tfrac{1}{2}. \quad (4.112)$$

We have therefore derived the inequalities to be satisfied by Poisson's ratio. The upper limit of a half, obviously, corresponds to the incompressibility ($\Delta V/V = 0$) of the material. No material with negative ν is known to exist. In reality, for most materials, ν falls between a quarter and a third. The values of ν are about 0.5 for rubber, 0.33 for plastics, 0.3 for metals and about 0.1 for ceramics. From the body of literature that exists on Poisson's ratio, special attention is drawn to the work of Köster and Franz (1961) who have reviewed the subject matter in detail.

4.10 Elastic limit and the concept of strength of materials

In the foregoing mathematical discussion on the stress–strain relation we were concerned with only small deformations in the sense that their products and squares were neglected. Now we conclude this section by summarising very briefly the experimental results which appear under more general conditions.

A schematic representation of the stress–strain relation found in a simple tension experiment is shown in figure 4.4. The first part of the graph OA represents the *Hookean* region where a linear relationship between stress and strain holds perfectly well. The point A is called the *limit of proportionality*. Thereafter, we meet a small region AB where Hooke's law of proportionality does not hold. The deformation in this region, however, disappears completely on the removal of stress. The point B on the diagram is designated as an *elastic limit*. The stress or strain corresponding to B is a quantitative measure of the elastic limit. It may be noted that the positions of A and B vary from material to material. For many systems A and B coincide and for steel B lies slightly below the *limit of proportionality*.

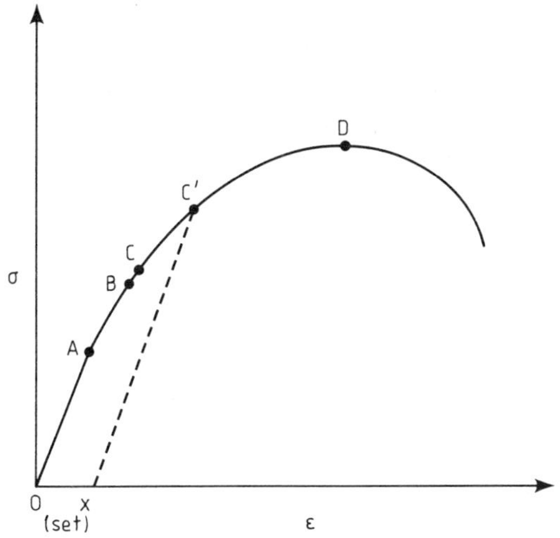

Figure 4.4 A schematic representation of the stress (σ) and strain (ε) variation observed in a tensile test experiment.

As the stress increases further we enter the region of *plasticity* where the curve is generally concave downwards. In this region for a given load the strain increases faster than in the elastic region and permanent deformations occur. Even after the complete removal of stress the specimen is left with a permanent deformation called a *set* which is depicted as Ox in figure 4.4. If the force required to produce a given set is small the material is said to be *soft* or *ductile* and otherwise *hard*. The stress at the beginning of the *plastic* region (see yield point C in the diagram) is called the *yield stress* or the *tensile yield strength* for which a finite amount of *set* has occurred. The process through which the yield stress is increased is called *work hardening* or *strain hardening*.

It may be noted, however, that the plastic flow in a material can be rather more easily initiated by applying *shear stress* than *tensile stress*. The shearing stress corresponding to the yield point C is called the *critical shear stress* and is much lower in magnitude than the tensile yield strength. The values of critical stress are typically of the order of $10^{-3}G$ which in hardened materials rise up to $G/30$. Why materials yield at such a low shearing stress, is a subject matter in its own right called dislocation theory. This is an exciting topic but falls beyond the scope of our introductory book. On this subject the books by Hirth and Lothe (1968), Nabarro (1967), Cottrell (1953) and Read (1953) may be found useful for detailed study.

Finally, we reach the point D on the stress–strain diagram which has a maximum value of stress called the *tensile strength*. This is often referred to as the *strength of the material*. Beyond this point the specimen begins to thin down at a particular cross section which is known as *necking*. Ultimately the specimen fractures at a stress less than the maximum which is called the *breaking stress*.

There are many known engineering methods through which a material can be further strengthened or values of the tensile strength increased. These processes include alloying, work hardening, annealing and cover many other methods leading to defect formation. Again, this is a vast topic in metallurgy and cannot be pursued any further here.

Problems

4.1 Suppose a solid body is subjected to stress with components

$$\sigma_{ij}=\begin{pmatrix} 5 & 3 & 0 \\ 3 & -3 & 0 \\ 0 & 0 & 7 \end{pmatrix} \times 10^8 \text{ dyn cm}^{-2} \qquad \text{(P4.1)}$$

(i) Determine the principal values and principal axes of the stress tensor.
(ii) What is the mean pressure on the solid due to the above stress tensor?

4.2 If the above solid is a cubic crystal with elastic constants (in the Voigt notation)

$$c_{11}=12\times 10^{11} \text{ dyn cm}^{-2}$$

$$c_{12}=6\times 10^{11} \text{ dyn cm}^{-2}$$

$$c_{44}=3\times 10^{11} \text{ dyn cm}^{-2}$$

calculate the strain components for the stress (P4.1) and express these both in the Voigt and in the tensor notation. Is the above cubic crystal elastically isotropic?

4.3 Referred to the principal axes a stress tensor has components (in units of dyn cm^{-2})

$$\sigma_1 = 5 \times 10^6 \qquad \sigma_2 = 5 \times 10^6 \qquad \sigma_3 = 12 \times 10^6.$$

Consider an element of surface whose normal $\boldsymbol{n}$ makes an angle of 30° with the x_3-axis and the projection of $\boldsymbol{n}$ on the x_1–x_2 plane makes an angle of 40° with the x_1-axis. Determine the value of the stress component σ_{nn}.

4.4 Show that for an elastically isotropic solid, the principal axes of the stress tensor are the same as those of the corresponding strain tensor.

4.5 Determine the stress–strain relations for an elastically isotropic solid in cylindrical polar coordinates (the relations between σ_{rr} and ε_{rr} etc).

4.6 In a long rod (made of elastically isotropic material) stress is applied in such a way that the non-vanishing strain component is ε_{33}, the x_3-axis being along the length of the rod.

(i) What are the non-vanishing components of the stress?

(ii) What is the ratio of the lateral to the longitudinal stress? Express your result in terms of Young's modulus and Poisson's ratio.

5

EQUILIBRIUM EQUATION AND ITS APPLICATION

The general form of the equilibrium equation derived in the previous Chapter shall be specialised here to an elastically isotropic body with an application to calculating the stress distribution in a long cylindrical pipe.

5.1 Equilibrium equations for an elastically isotropic body

We recall from the preceding Chapter that if a body is acted upon by the body forces F_i and the stress distribution σ_{ij} then the total force $\mathscr{F}^{(t)}$ acting per unit volume is

$$\mathscr{F}_i^{(t)} = \frac{\partial \sigma_{ij}}{\partial x_j} + \rho F_i \qquad i, j = 1, 2, 3. \tag{5.1}$$

Obviously for equilibrium the total force is zero and the above equation becomes the same as in (4.8). For an elastically isotropic body σ_{ij} can be written as (see equation (4.65)),

$$\sigma_{ij} = \lambda\, \Delta \delta_{ij} + 2\mu \varepsilon_{ij}.$$

Using this in equation (5.1) and considering $i = 1$, one has

$$\mathscr{F}_1^{(t)} = \lambda \frac{\partial(\Delta)}{\partial x_1} + 2\mu \frac{\partial \varepsilon_{11}}{\partial x_1} + 2\mu \frac{\partial \varepsilon_{21}}{\partial x_2} + 2\mu \frac{\partial \varepsilon_{31}}{\partial x_3} + \rho F_1. \tag{5.2}$$

Making substitutions for $\varepsilon_{11}, \varepsilon_{21}, \ldots$ from equation (3.12) it becomes

$$\mathscr{F}_1^{(t)} = (\lambda + \mu) \frac{\partial\, \Delta}{\partial x_1} + \mu \nabla^2 s_1 + \rho F_1. \tag{5.3a}$$

Similarly for $i = 2$ and $i = 3$, one obtains

$$\mathscr{F}_2^{(t)} = (\lambda + \mu) \frac{\partial\, \Delta}{\partial x_2} + \mu \nabla^2 s_2 + \rho F_2 \tag{5.3b}$$

$$\mathscr{F}_3^{(t)} = (\lambda + \mu) \frac{\partial\, \Delta}{\partial x_3} + \mu \nabla^2 s_3 + \rho F_3 \tag{5.3c}$$

or, in vector notation

$$\boldsymbol{\mathscr{F}}^{(t)} = (\lambda + \mu)\, \text{grad div}\, \boldsymbol{s} + \mu \nabla^2 \boldsymbol{s} + \rho \boldsymbol{F}. \tag{5.4}$$

Now if we use the vector identity

$$\text{grad div} = \text{curl curl} + \nabla^2 \tag{5.5}$$

equation (5.4) becomes

$$\mathscr{F}^{(t)} = (\lambda + 2\mu) \text{ grad div } \boldsymbol{s} - \mu \text{ curl curl } \boldsymbol{s} + \rho \boldsymbol{F}. \tag{5.6}$$

For equilibrium ($\mathscr{F}^{(t)} = 0$) and hence the displacements must satisfy

$$(\lambda + 2\mu) \text{ grad div } \boldsymbol{s} - \mu \text{ curl curl } \boldsymbol{s} + \rho \boldsymbol{F} = 0 \tag{5.7}$$

which are the desired equations of equilibrium for isotropic bodies. The last term, $\rho \boldsymbol{F}$, represents the body forces, and we recall that these are nothing but the manifestations of the gravitational forces. In elasticity problems the effect of body forces is usually neglected because it produces a negligibly small deformation in comparison with other external or surface forces. To visualise this, consider a uniform wire of Cu, say, 100 cm long. The stress acting on this due to the earth's gravity ($\sigma \simeq F/A = \rho g l$, where g is the acceleration due to gravity) is of the order of 10^5 dyn cm^{-2}. However, to produce a deformation of 0.1%, one requires $\sigma = E(\Delta l / l) \simeq 10^8$ dyn cm^{-2}. Thus the effect of body forces could be neglected† and in that case the equilibrium equation becomes

$$(\lambda + 2\mu) \text{ grad div } \boldsymbol{s} - \mu \text{ curl curl } \boldsymbol{s} = 0. \tag{5.8}$$

Equations (5.7) or (5.8), together with the appropriate boundary conditions on σ_{ij} allow one to estimate the stress distribution in various structures like bridges, aeroplane wings, pipes at high pressures, etc. In the following section, we shall apply the equilibrium equation as derived here to calculate the stress distribution in a long cylindrical pipe. This problem has obvious practical relevance for canning, the flow of oil and gas in pipes at high pressure p and in many other similar problems where one must know that the pipe withstands the pressure and does not buckle under it.

5.2 Stress in a long cylindrical pipe

Consider a long cylindrical pipe (see figure 5.1) having internal and external radii R_1 and R_2, respectively, which is filled with fluid at pressure p. Our aim here is to determine the stresses in the material of the pipe for a given value of p. We assume that the pipe is long enough so that end effects can be neglected and also that $\boldsymbol{F} = 0$. The boundary conditions‡ for σ_{ij} are

$$\begin{aligned} \sigma_{rr} &= -p \qquad && \text{at } r = R_1 \\ &= 0 \qquad && \text{at } r = R_2 \end{aligned} \tag{5.9}$$

† For huge structures like tall buildings, bridges, etc, the body forces cannot be neglected.

‡ Here the effect of atmospheric pressure has been neglected.

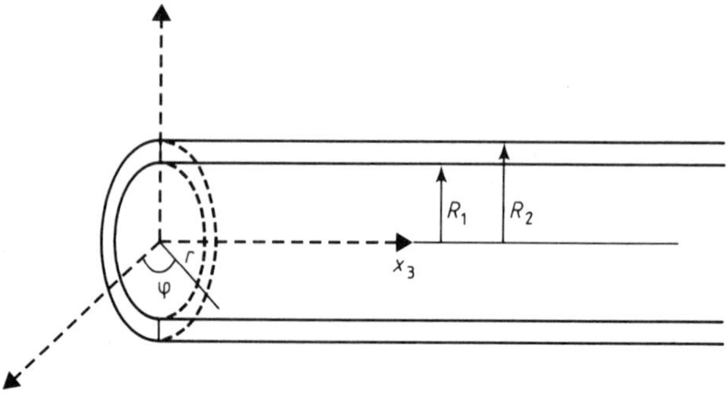

Figure 5.1 A long cylindrical pipe filled with fluid at pressure p is shown.

and all other components are zero. From the symmetry of the problem it is necessary that the displacements $\boldsymbol{s}$ must be along r and a function of r only, so that

$$s_r(r) \neq 0 \qquad \text{but } s_\varphi = s_z = 0. \tag{5.10}$$

Making use of (5.10) in the expressions for curl and div in cylindrical polar coordinates (see equations (P2.3) through (P2.6)) one has

$$\text{curl } \boldsymbol{s} = 0 \qquad \text{and div } \boldsymbol{s} = \Delta = \frac{1}{r}\frac{\partial}{\partial r}(rs_r) \tag{5.11}$$

Hence the equilibrium equation (5.8) becomes

$$(\lambda + 2\mu) \text{ grad div } \boldsymbol{s} = 0$$

or

$$\text{div } \boldsymbol{s} = \text{constant.} \tag{5.12}$$

For later convenience let us take

$$\text{div } \boldsymbol{s} = 2a \tag{5.13}$$

where a is a constant. This in conjunction with equation (5.11) yields

$$\partial(rs_r)/\partial r = 2ar. \tag{5.14}$$

Now, by integrating both sides, we obtain

$$s_r = ar + br^{-1} \tag{5.15}$$

where b is another constant. Thus, the strain components become (see equations (3.82) through (3.87))

$$\varepsilon_{rr} = \frac{\partial s_r}{\partial r} = a - \frac{b}{r^2} \tag{5.16a}$$

$$\varepsilon_{\varphi\varphi}=\frac{s_r}{r}+\frac{1}{r}\frac{\partial s_\varphi}{\partial\varphi}=a+\frac{b}{r^2} \tag{5.16b}$$

and all other $\varepsilon_{ij}=0$. The dilatation is therefore

$$\Delta=\varepsilon_{rr}+\varepsilon_{\varphi\varphi}+\varepsilon_{zz}=2a. \tag{5.16c}$$

Now the stress–strain relation for isotropic materials in cylindrical coordinates can be written as

$$\sigma_{rr}=\lambda\,\Delta+2\mu\varepsilon_{rr}=2a(\lambda+\mu)-2\mu br^{-2} \tag{5.17a}$$

$$\sigma_{\varphi\varphi}=\lambda\,\Delta+2\mu\varepsilon_{\varphi\varphi}=2a(\lambda+\mu)+2\mu br^{-2} \tag{5.17b}$$

$$\sigma_{zz}=\lambda\,\Delta=2a\lambda \tag{5.17c}$$

$$\sigma_{r\varphi}=\sigma_{\varphi z}=\sigma_{rz}=0. \tag{5.17d}$$

where λ and μ are as usual Lame's constants. When equation (5.17*a*) is subjected to the boundary conditions (5.9) one readily obtains

$$\frac{b}{a}=R_2^2\left(\frac{\lambda+\mu}{\mu}\right) \tag{5.18}$$

and

$$-p=2a(\lambda+\mu)(1-R_2^2R_1^{-2}) \tag{5.19}$$

which on solving for a and b gives

$$a=\frac{p}{2}\frac{R_1^2}{(\lambda+\mu)(R_2^2-R_1^2)} \tag{5.20}$$

$$b=\frac{p}{2}\frac{R_1^2R_2^2}{\mu(R_2^2-R_1^2)}. \tag{5.21}$$

Making use of these constants in equations (5.16), the strain components become

$$\varepsilon_{rr}=\frac{p}{2}\frac{R_1^2}{(\lambda+\mu)(R_2^2-R_1^2)}\left[1-\left(\frac{\lambda+\mu}{\mu}\right)\frac{R_2^2}{r^2}\right] \tag{5.22a}$$

$$\varepsilon_{\varphi\varphi}=\frac{p}{2}\frac{R_1^2}{(\lambda+\mu)(R_2^2-R_1^2)}\left[1+\left(\frac{\lambda+\mu}{\mu}\right)\frac{R_2^2}{r^2}\right]. \tag{5.22b}$$

If we substitute ε_{rr}, $\varepsilon_{\varphi\varphi}$ and a back in equation (5.17), we obtain the following expressions for the stress components

$$\sigma_{rr}=\frac{pR_1^2}{(R_2^2-R_1^2)}\left(1-\frac{R_2^2}{r^2}\right) \tag{5.23a}$$

$$\sigma_{\varphi\varphi}=\frac{pR_1^2}{(R_2^2-R_1^2)}\left(1+\frac{R_2^2}{r^2}\right) \tag{5.23b}$$

$$\sigma_{zz}=\frac{p\lambda}{\lambda+\mu}\frac{R_1^2}{R_2^2-R_1^2}=\frac{2p\nu R_1^2}{R_2^2-R_1^2} \tag{5.23c}$$

$$\sigma_{r\varphi}=\sigma_{\varphi z}=\sigma_{rz}=0. \tag{5.23d}$$

We find that with a decrease in the thickness (R_2-R_1) of the material of the pipe the stresses along the r-, φ- and z-axes of the pipe increase for a given p. Furthermore, since σ_{rr}, $\sigma_{\varphi\varphi}$ and σ_{zz} are not all equal to each other there is a non-vanishing shearing stress across arbitrarily directed planes. The material would yield once the shearing stress on any plane becomes greater than the critical value. Hence, knowing the critical shear stress of the material, one can choose the appropriate thickness (R_2-R_1) which would sustain the desired pressure p. Finally, it may be remarked that the present solution is only an approximation to any real physical situation since end effects have been neglected here.

There are many other problems such as the stretching of a membrane, the twisting of rods, the bending of beams, spherical shell etc which are of great physical interest and are dealt with in different books on elasticity (see, for example, Landau and Lifshitz 1959a, Love 1944). We do not intend to pursue them here; rather we shall at this point go on to the equations of motion and wave propagation.

Problems

5.1 Using the equilibrium equations for a solid, show that the equilibrium equation for a fluid at rest and acted upon by a body force $\boldsymbol{F}$ per unit mass is

$$\text{grad}\, p-\rho\boldsymbol{F}=0 \tag{P5.1}$$

where p is the pressure and ρ is the density of the fluid.

5.2 Consider a long rod of elastically isotropic material of length l standing vertically in a vacuum in equilibrium under the gravitational field of the earth, then:

(i) What are the boundary conditions for σ_{ij} on the various surfaces of the rod?

(ii) Solve for σ_{ij}.

(iii) Determine the strain components ε_{ij}.

(iv) Solve for the displacement $\boldsymbol{s}$.

5.3 Use the equilibrium equation to show that in the absence of body forces, the dilatation Δ satisfies the differential equation

$$\frac{\partial^2\Delta}{\partial x_1^2}+\frac{\partial^2\Delta}{\partial x_2^2}+\frac{\partial^2\Delta}{\partial x_3^2}=0. \tag{P5.2}$$

5.4 A rod of length l made of an elastically isotropic material is hanging vertically under its own weight. Show that the following displacements maintain equilibrium:

$$s_1 = -\sigma\rho g x_1 x_3/E$$

$$s_2 = -\sigma\rho g x_2 x_3/E$$

$$s_3 = \frac{\rho g}{2E}(x_3^2 - l^2) + \frac{\sigma\rho g}{2E}(x_1^2 + x_2^2)$$

where σ is Poisson's ratio, E is Young's modulus and ρ is the density of the material.

6

EQUATION OF MOTION AND WAVE PROPAGATION

The previous Chapters were concerned with some of the elastic problems that were independent of time. We shall now describe the equation of motion. When an elastic body is subjected to forces which are suddenly applied or to variable forces, motion is produced. Initially the deformations are produced at the point of action and, in the course of time, propagate through the medium in the form of elastic waves. It is the purpose of this Chapter to set up the equation of motion which is then used to discuss the wave propagation through an elastically isotropic medium.

6.1 Velocity flow function and material derivative

In particle mechanics the motion of every particle is treated separately. Each particle is identified with its position coordinates $x_i\,(x_1, x_2, x_3)$ at time t. As the particle moves the coordinates x_i change with t, that is to say, x_i and t are dependent variables. A similar description for a continuous medium which consists of an infinite number of particles is possible but very complex. Instead, we shall see that it is easier to describe the motion of a continuous medium by introducing a function, $\boldsymbol{v}(\boldsymbol{x}, t)$, where $\boldsymbol{x}$ is now fixed in space. This is known as the spatial description of a continuous medium. $\boldsymbol{v}(\boldsymbol{x}, t)$ is called the velocity flow function or simply the velocity function which gives the instantaneous velocity of a particle or particles located at a point $\boldsymbol{x}(x_1, x_2, x_3)$ fixed in space at time t. Here $\boldsymbol{x}$ and t are independent variables.

The function $\boldsymbol{v}(\boldsymbol{x}, t)$ can be visualised by looking at the flow of a river or the flow in a pipe. A flowing particle which at time t is located at a point with coordinates $\boldsymbol{x}$ moves to another point with coordinates $x_i + v_i\,\mathrm{d}t$ in time $\mathrm{d}t$. If $v_i(\boldsymbol{x},t)$ and $v_i(x_i + v_i\mathrm{d}t,\ t+\mathrm{d}t)$ are the velocity functions at two points then the instantaneous acceleration $a_i(\boldsymbol{x}, t)$ is given by $(i = 1, 2, 3)$

$$a_i(\boldsymbol{x}, t) = \lim_{\mathrm{d}t \to 0} \frac{[v_i(x_1 + v_1\mathrm{d}t, x_2 + v_2\mathrm{d}t, x_3 + v_3\mathrm{d}t, t + \mathrm{d}t) - v_i(\boldsymbol{x}, t)]}{\mathrm{d}t}. \tag{6.1}$$

Making use of Taylor's expansion and omitting the higher-order terms, one obtains

$$a_i(\boldsymbol{x}, t) = \frac{\partial v_i}{\partial t} + v_1 \frac{\partial v_i}{\partial x_1} + v_2 \frac{\partial v_i}{\partial x_2} + v_3 \frac{\partial v_i}{\partial x_3} \tag{6.2}$$

or in compact form,

$$a_i(\boldsymbol{x}, t) = \frac{\partial v_i}{\partial t} + \sum_{j=1}^{3} v_j \frac{\partial v_i}{\partial x_j} \tag{6.3}$$

which is an expression for the particle or material acceleration in a continuous medium. The right-hand side of (6.3) consists of two terms, the first denotes the rate of change of v_i with respect to time at a fixed point in space and is therefore called the local acceleration. The second term, on the other hand, represents the rate of change of v_i due to the change of the position of the particle from one point to another in a fixed time t. We further emphasise that the argument given here may be extended to any physical quantity associated with particles. If we denote any such quantity by a general function $\psi(\boldsymbol{x}, t)$ then we can write

$$\lim_{dt \to 0} \frac{1}{dt} [\psi(x_i + v_i\, dt, t + dt) - \psi(\boldsymbol{x}, t)] = \frac{\partial \psi}{\partial t} + \sum_j v_j \frac{\partial \psi}{\partial x_j}. \tag{6.4}$$

Hence we define a material derivative,

$$\frac{D}{Dt} = \frac{\partial}{\partial t} + \sum_j v_j \frac{\partial}{\partial x_j} \tag{6.5}$$

which will prove very useful in calculating the rate of change of a quantity associated with particle or particles in motion. For example, taking the displacement vector $\boldsymbol{s}(\boldsymbol{x}, t)$ it can easily be used to obtain the material velocity,

$$v_i(\boldsymbol{x}, t) = \frac{Ds_i(\boldsymbol{x}, t)}{Dt} = \frac{\partial s_i}{\partial t} + \sum_j v_j \frac{\partial s_i}{\partial x_j} \tag{6.6}$$

and similarly taking $v_i(\boldsymbol{x}, t)$ one obtains an expression for the material acceleration which is the same as in equation (6.3). Hereafter we shall use v_i for $v_i(\boldsymbol{x}, t)$, s_i for $s_i(\boldsymbol{x}, t)$ and similarly for the other quantities.

6.2 The equation of motion

When the right-hand side of the equilibrium equation (4.8) which is zero is replaced by inertial forces, we obtain the equations of motion. Consider an elastic material of density ρ ($\equiv \rho(x, t)$) enclosed in a small volume d^3x ($= dx_1\, dx_2\, dx_3$) then the equations of motion are

$$(\rho\, d^3x) a_i = \left(\rho F_i + \sum_j \frac{\partial \sigma_{ji}}{\partial x_j} \right) d^3x \qquad i, j = 1, 2, 3 \tag{6.7}$$

or in terms of the material derivative

$$\rho \frac{Dv_i}{Dt} = \rho F_i + \sum_j \frac{\partial \sigma_{ji}}{\partial x_j}. \tag{6.8}$$

It may be noticed that up to this point in the derivation of (6.8) no assumption has been made regarding σ_{ij} and F_i. Therefore, equation (6.8) constitutes an exact form of the equation of motion for an elastic body. We shall, however, see that it simplifies considerably subject to boundary conditions and assumptions made depending upon the nature of the problem.

If we consider an ideal fluid (inviscid and incompressible)

$$\sigma_{ij} = -p\delta_{ij}$$

then equation (6.8) becomes:

$$\rho \frac{\mathrm{D}v_i}{\mathrm{D}t} = \rho F_i - \frac{\partial p}{\partial x_i} \tag{6.9}$$

which is the celebrated Euler equation of motion, one of the fundamental equations of fluid mechanics. Observe that both equations (6.8) and (6.9) are non-linear. This makes the solution difficult, particularly for solids and therefore one resorts to some reasonable approximations.

6.3 The linear approximation for solids

To begin with we recapitulate some of the approximations we have already made regarding σ_{ij} and ε_{ij} such as, (i) σ_{ij} and ε_{ij} are related linearly through Hooke's law and (ii) the higher powers of $\partial s_i/\partial x_j$ are neglected in defining ε_{ij}. Likewise we further assume that

$$v_i = \frac{\mathrm{D}s_i}{\mathrm{D}t} = \frac{\partial s_i}{\partial t} + v_j \frac{\partial s_i}{\partial x_j}$$

$$\simeq \frac{\partial s_i}{\partial t} \tag{6.10}$$

and

$$\frac{\mathrm{D}v_i}{\mathrm{D}t} = \frac{\partial v_i}{\partial t} + v_j \frac{\partial v_i}{\partial x_j}$$

$$\simeq \frac{\partial v_i}{\partial t} = \frac{\partial}{\partial t}\left(\frac{\partial s_i}{\partial t}\right). \tag{6.11}$$

Hence, the equation of motion (6.8) becomes

$$\rho \frac{\partial^2 s_i}{\partial t^2} = \rho F_i + \sum_j \frac{\partial \sigma_{ij}}{\partial x_j}. \tag{6.12}$$

Moreover, in accordance with our assumption of retaining only linear terms in $\mathbf{s}$, the density ρ may be replaced by its average value ρ_0. Replacing† ρ

† Since $\rho V = \rho_0 V_0$ or $\rho V_0(1 + \operatorname{div} \mathbf{s}) = \rho_0 V_0$.

($=\rho_0(1+\text{div}\ \boldsymbol{s})^{-1}$) by ρ_0 and substituting for the right-hand-side terms in (6.12) as in equations (5.1) and (5.4) one obtains

$$\rho_0 \frac{\partial^2 \boldsymbol{s}}{\partial t^2} = \rho \boldsymbol{F} + (\lambda + \mu)\ \text{grad div}\ \boldsymbol{s} + \mu \nabla^2 \boldsymbol{s}. \tag{6.13}$$

In the absence of body forces ($\boldsymbol{F}=0$),

$$\rho_0 \frac{\partial^2 \boldsymbol{s}}{\partial t^2} = (\lambda + \mu)\ \text{grad div}\ \boldsymbol{s} + \mu \nabla^2 \boldsymbol{s}. \tag{6.14}$$

These equations are immensely helpful in studying the vibrations and wave propagation in solids using the appropriate boundary conditions.

6.4 Wave propagation in an infinite isotropic material

We shall discuss here the solution of equation (6.14) for an unbounded isotropic medium. Some of the results which emerge under different conditions are enumerated below.

6.4.1 Shear waves

Assume first that the deformation produced in the medium is such that no change in volume occurs, i.e. div $\boldsymbol{s}=0$. The equation (6.14) becomes

$$\frac{\partial^2 \boldsymbol{s}}{\partial t^2} = \frac{\mu}{\rho_0} \nabla^2 \boldsymbol{s}. \tag{6.15}$$

This is an equation for waves called isovoluminous or shear waves. The velocity of propagation is

$$C_t = (\mu/\rho_0)^{1/2}. \tag{6.16}$$

6.4.2 Rotations associated with arbitrary $\boldsymbol{s}$

If one takes the curl of each term in equation (6.14) and uses the identity curl grad $=0$, one obtains

$$\frac{\partial^2}{\partial t^2}\ \text{curl}\ \boldsymbol{s} = \frac{\mu}{\rho_0} \nabla^2\ \text{curl}\ \boldsymbol{s} \tag{6.17}$$

or in terms of $\boldsymbol{R}$ ($\equiv \frac{1}{2}$ curl $\boldsymbol{s}$) the wave equation becomes

$$\frac{\partial^2 \boldsymbol{R}}{\partial t^2} = \frac{\mu}{\rho_0} \nabla^2 \boldsymbol{R} \tag{6.18}$$

Thus the velocity of propagation is the same as in equation (6.16).

6.4.3 Irrotational or dilatational waves

Consider now the case when the deformation produced by the waves is not accompanied by rotation, i.e. curl $\boldsymbol{s}=0$. If we use the vector identity

$$\text{grad div} = \text{curl curl} + \nabla^2$$

then equation (6.14) becomes

$$\frac{\partial^2 \boldsymbol{s}}{\partial t^2} = \frac{(\lambda + 2\mu)}{\rho_0} \nabla^2 \boldsymbol{s}. \tag{6.19}$$

This is an equation for irrotational waves. The velocity of propagation is

$$C_l = [(\lambda + 2\mu)/\rho_0]^{1/2}. \tag{6.20}$$

If λ and μ are expressed in terms of elastic moduli (see §4.8) then

$$C_l = [(B + \tfrac{4}{3}G)/\rho_0]^{1/2}. \tag{6.21}$$

On the other hand, if we take the divergence of equation (6.19) then one has

$$\frac{\partial^2 \Delta}{\partial t^2} = \left(\frac{\lambda + 2\mu}{\rho_0}\right) \nabla^2 \Delta. \tag{6.22}$$

The dilatation Δ (= div $\boldsymbol{s}$) thus satisfies the wave equation which has the same velocity of propagation as that of irrotational waves. For this reason the irrotational waves are also called dilatational waves. This name is frequently used in the literature. It may, however, be noted that the deformation produced by the waves is not just a pure dilatation because the velocity expression contains both the bulk modulus and the shear modulus.

6.4.4 Properties of irrotational waves and shear waves

We have proof that irrotational waves and shear waves are the only types of wave possible and that they propagate independently of one another.

We shall arrive at this result by employing two important results of vector analysis, namely (i) a continuous vector field function $\boldsymbol{A}$ can be uniquely (apart from additive constants) represented as the sum of an irrotational field function $\boldsymbol{A}_l$ and a solenoidal field function $\boldsymbol{A}_t$,

$$\boldsymbol{A} = \boldsymbol{A}_l + \boldsymbol{A}_t \tag{6.23a}$$

where

$$\text{curl } \boldsymbol{A}_l = 0 \qquad \text{div } \boldsymbol{A}_t = 0 \tag{6.23b}$$

and (ii) if both the curl and divergence of a vector vanish everywhere in space, the vector may be taken to be zero.

In our case, following (i), the displacement $\boldsymbol{s}$ can be written as

$$\boldsymbol{s} = \boldsymbol{s}_l + \boldsymbol{s}_t \tag{6.24a}$$

such that

$$\text{curl } s_l = 0 \qquad \text{div } s_t = 0. \tag{6.24b}$$

Now substituting equation (6.24) into (6.14) and using the vector analysis result (ii) one obtains

$$\left(\frac{\partial^2}{\partial t^2} - \frac{\lambda + 2\mu}{\rho_0}\nabla^2\right) s_l = 0 \tag{6.25a}$$

$$\left(\frac{\partial^2}{\partial t^2} - \frac{\mu}{\rho_0}\nabla^2\right) s_t = 0 \tag{6.25b}$$

where the first equation represents the propagation of irrotational waves and the second represents the propagation of shear waves. Thus, in an infinite isotropic medium equation (6.14) is equivalent to two independent wave equations (6.25*a*) and (6.25*b*). Obviously the velocity of propagation of the respective waves are C_l and C_t and are independent of one another. We have as an example, for instance, seismic disturbances which produce both irrotational and shear waves which propagate independently of one another.

6.4.5 *The displacement* **s** *in plane waves*

The waves at a greater distance from the source can be considered as planc waves. These are a simple type of wave but are very useful in understanding the displacements in the medium away from the point of disturbance. For a plane wave propagating in the direction $\boldsymbol{k}$, let

$$\boldsymbol{s} = s_0 \boldsymbol{e} \exp\left[\mathrm{i}(\omega t - \boldsymbol{k}\cdot\boldsymbol{x})\right] \tag{6.26}$$

where $\boldsymbol{e}$ is the unit vector along $\boldsymbol{s}$, ω is the frequency and the wavenumber $k = 2\pi/\lambda$.

Consider, first, the shear waves, i.e.

$$\text{div } \boldsymbol{s} = \frac{\partial s_1}{\partial x_1} + \frac{\partial s_2}{\partial x_2} + \frac{\partial s_3}{\partial x_3} = 0.$$

This equation, when used in conjunction with equation (6.26) implies that

$$\boldsymbol{e}\cdot\boldsymbol{k} = 0 \qquad \text{or } \boldsymbol{e}\perp\boldsymbol{k} \tag{6.27}$$

i.e. the displacement in the medium is perpendicular to the direction of propagation of the waves. Thus plane shear waves are transverse waves. Irrotational waves, curl $\boldsymbol{s} = 0$, on the other hand, imply that

$$\boldsymbol{e}\times\boldsymbol{k} = 0 \qquad \boldsymbol{e}\parallel\boldsymbol{k}. \tag{6.28}$$

The displacements in the medium occur along the direction of propagation of the waves. Therefore, plane irrotational waves are called longitudinal waves. C_l and C_t, accordingly, are longitudinal and transverse velocities. In seismology irrotational waves are also known as P waves (P stands for pressure) and shear waves are known as S waves.

6.4.6 Velocity and spectrum of elastic waves

Making use of equations (6.16) and (6.20) one has

$$\frac{C_l}{C_t}=\left(\frac{\lambda+2\mu}{\mu}\right)^{1/2} \tag{6.29}$$

or in terms of Poisson's ratio v (see equation (4.89))

$$\frac{C_l}{C_t}=\left(\frac{2-2v}{1-2v}\right)^{1/2}. \tag{6.30}$$

Thus depending upon v, the ratio C_l/C_t is different for different materials. In particular for $v=\frac{1}{4}$,

$$C_l/C_t=\sqrt{3}. \tag{6.31}$$

Table 6.1 lists values of C_l and C_t for some solid materials.

In a medium which cannot sustain shear stress ($\mu\sim 0$), for example ideal fluids and gases, no shear waves can be propagated. The velocity of propagation of a longitudinal wave is simply (see equation (6.21))

$$C_l=(B/\rho_0)^{1/2}. \tag{6.32}$$

Typical values of C_l in a liquid are of the order of 1200 m s^{-1} and that in air is 300 m s^{-1}.

Table 6.1 Values of C_l and C_t (both in units of 10^3 m s^{-1}) of elastic waves in different systems at room temperature†.

	C_l	C_t	C_l/C_t
Potassium	2.67	1.45	1.84
Ice (257 K)	3.58	1.83	1.95
Lead	2.23	0.95	2.34
Sodium chloride	4.46	2.58	1.73
Aluminum	6.45	3.12	2.06
Quartz	6.16	4.19	1.47
Copper	4.69	2.31	2.03
Iron	5.95	3.29	1.81
Beryllium	13.28	8.74	1.52
Diamond	17.26	11.59	1.49

† Values are taken from the compilation by Simmons and Wang (1971).

The frequencies of elastic waves, however, cover a wide spectral range. Longitudinal elastic waves audible to the human ear are called sound waves or sonic waves, the frequency lying in the range from 15 Hz to about 20 000 Hz. Waves with frequencies below the audible range are called infrasonic waves and those above are called ultrasonic waves. Finally the vibration (of thermal origin) of atoms in a solid can also be regarded as elastic waves. The quantum of such a wave is called the phonon. The frequency ranges from 0 to about 10^{13} Hz.

6.5 Elastic waves in a finite medium

In the preceding section we applied the equation of motion to study the propagation of elastic waves in an infinite isotropic medium (unbounded) and we saw that only irrotational and shear waves can be propagated. In finite media, however, the situation is quite different and the propagation of elastic waves depends very much on the nature of the boundary conditions. We shall describe this very briefly by taking as a simple example the propagation of longitudinal waves in a long thin rod.

Consider a long thin rod whose diameter is negligibly small in comparison with its length ($r_0 \ll l$). Let one of its ends be subject to a periodic longitudinal stress σ_{11} ($=\sigma_1$). Since by assumption the thickness of the rod is negligibly small all other components of σ_{ij} may be assumed to be zero. Therefore the relevant longitudinal strain is

$$\varepsilon_1 = \partial s_1/\partial x_1 = s_{11}\sigma_1. \tag{6.33}$$

In the absence of body forces the equation of motion along the x_1-direction becomes

$$\begin{aligned} \rho_0 \frac{\partial^2 s_1}{\partial t^2} &= \frac{\partial \sigma_{ji}}{\partial x_j} \\ &= \frac{\partial \sigma_1}{\partial x_1}. \end{aligned} \tag{6.34}$$

Substituting for σ_1 from equation (6.33),

$$\frac{\partial^2 s_1}{\partial t^2} = \frac{1}{\rho_0 s_{11}} \frac{\partial^2 s_1}{\partial x_1^2}. \tag{6.35}$$

This represents the propagation of longitudinal waves in a thin rod. The wave velocity is

$$C = \left(\frac{1}{\rho_0 s_{11}}\right)^{1/2} = \left(\frac{E}{\rho_0}\right)^{1/2} \tag{6.36}$$

which is obviously different from either of the irrotational or shear waves in an infinite medium. Equation (6.36) has been obtained assuming a rod of negligible thickness and therefore, after making corrections for finite transverse size (r_0), one obtains (see, for example, Love (1944) p 290, Mason (1958) p 42),

$$C=(E/\rho_0)^{1/2}(1-\tfrac{1}{4}v^2k^2r_0^2). \tag{6.37}$$

Finally, we wish to point out that the subject matter of the present section only provides a glimpse of the true picture. Indeed it is much more interesting and involved than it appears here. Mention may be made of other important waves such as torsional, flexural as well as surface waves which appear in bounded media. Among these, surface waves are of particular importance in seismology. Surface waves are also known as Rayleigh waves. For a detailed mathematical derivation a more advanced text such as that of Love (1944) should be consulted.

6.6 Reflection and refraction of plane elastic waves

When an elastic wave meets a boundary surface between two media, like other waves, it is also reflected as well as refracted. A characteristic feature is that longitudinal waves in general are not only reflected (or refracted) as longitudinal waves but also as transverse waves, and similarly transverse waves are reflected or refracted as both transverse and longitudinal waves. This phenomenon is known as mode conversion. We first address here a relatively simple question as to how and why this mode conversion occurs.

6.6.1 Reflection of longitudinal waves from an isotropic-medium–vacuum interface

Consider an elastic medium in the half space to the left of the plane $x_1=0$. To the right we assume there is a vacuum. A plane longitudinal wave of amplitude A_1 which is propagating in the x_1–x_2 plane with velocity $C_l=\omega/k_l$ meets the surface of separation (figure 6.1) at an angle of incidence α,

$$s_l^{\text{inc}}=\boldsymbol{e}A_1\exp[\mathrm{i}(\omega t-\boldsymbol{k}_l\cdot\boldsymbol{x})] \tag{6.38}$$

where $\boldsymbol{e}$ $(\cos\alpha, \sin\alpha, 0)$ is the unit vector along $\boldsymbol{s}$ and

$$\boldsymbol{k}_l\cdot\boldsymbol{x}=k_lx_1\cos\alpha+k_lx_2\sin\alpha.$$

Making use of the abbreviation

$$\psi_1=A_1\exp\{\mathrm{i}[\omega t-(k_lx_1\cos\alpha+k_lx_2\sin\alpha)]\} \tag{6.39}$$

equation (6.38) becomes

$$s_l^{\text{inc}}=\boldsymbol{e}\psi_1. \tag{6.40}$$

Let us now frame the boundary conditions for our problem. We note that to

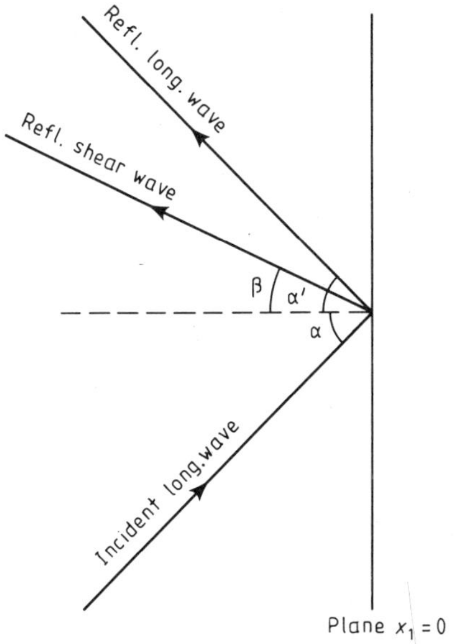

Figure 6.1 The reflection of a longitudinal wave at an isotropic-solid–air interface as shear waves and longitudinal waves.

the right of the interface is a vacuum and hence all the stress components are zero and so the same must be true to the left. The boundary conditions for stress components are, therefore,

$$\sigma_{11} = \sigma_{12} = \sigma_{13} = 0 \qquad \text{at } x_1 = 0. \tag{6.41}$$

This as we shall see presently can only be satisfied if both longitudinal and transverse waves are reflected. Anticipating this is the case let the longitudinal wave component be reflected at an angle α'. If the angle of reflection for the transverse component is β and the velocity of propagation is $C_t = \omega/k_t$ then as before one may write

$$\boldsymbol{s}_l^{\text{ref}} = \boldsymbol{e}' A_2 \exp\{i[\omega t - k_l(-x_1 \cos\alpha' + x_2 \sin\alpha')]\} \tag{6.42}$$

$$\boldsymbol{s}_t^{\text{ref}} = \boldsymbol{e}'' A_3 \exp\{i[\omega t - k_t(-x_1 \cos\beta + x_2 \sin\beta)]\} \tag{6.43}$$

where A_2, A_3 are amplitudes of the reflected longitudinal and the reflected transverse waves respectively. $\boldsymbol{e}'(-\cos\alpha', \sin\alpha', 0)$ and $\boldsymbol{e}''(\sin\beta, \cos\beta, 0)$ are unit vectors for the two displacements. It may be noted that the vibrations in the shear wave have only been considered in the x_1x_2 plane because vibrations parallel to the x_3-direction are absent by virtue of the boundary condition, namely $\sigma_{13} = 0$. Writing

$$\psi_2 = A_2 \exp\{i[\omega t - k_l(-x_1 \cos\alpha' + x_2 \sin\alpha')]\} \tag{6.44}$$

$$\psi_3 = A_3 \exp\{i[\omega t - k_t(-x_1 \cos\beta + x_2 \sin\beta)]\} \tag{6.45}$$

equations (6.42) and (6.43) become

$$\boldsymbol{s}_l^{\text{ref}} = \boldsymbol{e}'\psi_2 \tag{6.46}$$

$$\boldsymbol{s}_t^{\text{ref}} = \boldsymbol{e}''\psi_3. \tag{6.47}$$

The total displacement in the medium is, therefore,

$$\begin{aligned} \boldsymbol{s} &= \boldsymbol{s}_l^{\text{inc}} + \boldsymbol{s}_l^{\text{ref}} + \boldsymbol{s}_t^{\text{ref}} \\ &= \boldsymbol{e}\psi_1 + \boldsymbol{e}'\psi_2 + \boldsymbol{e}''\psi_3. \end{aligned} \tag{6.48}$$

Having defined $\boldsymbol{s}$ in the medium, it is now straightforward to obtain the strain components which, in turn, will allow us to find the stress components σ_{12} and σ_{11}. We first write

$$\begin{aligned} \varepsilon_{11} &= \frac{\partial s_1}{\partial x_1} \\ &= e_1 \frac{\partial \psi_1}{\partial x_1} + e_1' \frac{\partial \psi_2}{\partial x_1} + e_1'' \frac{\partial \psi_3}{\partial x_1} \end{aligned} \tag{6.49}$$

where $e_1 = \cos\alpha$, $e_1' = -\cos\alpha'$ and $e_1'' = \sin\beta$. Using equations (6.39), (6.44) and (6.45) this becomes

$$\varepsilon_{11} = -\mathrm{i}(k_l\psi_1 \cos^2\alpha + k_l\psi_2 \cos^2\alpha' - k_t\psi_3 \sin\beta \cos\beta). \tag{6.50}$$

Similarly for ε_{22},

$$\varepsilon_{22} = \frac{\partial s_2}{\partial x_2} = e_2 \frac{\partial \psi_1}{\partial x_2} + e_2' \frac{\partial \psi_2}{\partial x_2} + e_2'' \frac{\partial \psi_3}{\partial x_2} \tag{6.51}$$

where $e_2 = \sin\alpha$, $e_2' = \sin\alpha'$ and $e_2'' = \cos\beta$. On using the explicit forms of ψ_1, ψ_2, ψ_3 this reduces to

$$\varepsilon_{22} = -\mathrm{i}(k_l\psi_1 \sin^2\alpha + k_l\psi_2 \sin^2\alpha' + k_t\psi_3 \sin\beta \cos\beta). \tag{6.52}$$

For the strain components, ε_{12}, one has

$$\begin{aligned} \varepsilon_{12} &= \frac{1}{2}\left(\frac{\partial s_1}{\partial x_2} + \frac{\partial s_2}{\partial x_1}\right) \\ &= \frac{1}{2}\left(\frac{\partial}{\partial x_2}(e_1\psi_1 + e_1'\psi_2 + e_1''\psi_3) + \frac{\partial}{\partial x_1}(e_2\psi_1 + e_2'\psi_2 + e_2''\psi_3)\right) \\ &= \tfrac{1}{2}(-\mathrm{i})(k_l\psi_1 \sin 2\alpha - k_l\psi_2 \sin 2\alpha' - k_t\psi_3 \cos 2\beta). \end{aligned} \tag{6.53}$$

We now return to the boundary condition (6.41). The boundary condition $\sigma_{13} = 0$ is automatically satisfied because no motion along the x_3-direction has been considered. The condition for σ_{12} and σ_{11}, however, can be written as

$(\sigma_{ij}=2\mu\varepsilon_{ij}+\lambda\,\Delta\delta_{ij})$:

$$\sigma_{12}=2\mu\varepsilon_{12}=0 \qquad \text{at } x=0 \tag{6.54a}$$

$$\sigma_{11}=(\lambda+2\mu)\varepsilon_{11}+\lambda\varepsilon_{22}=0 \tag{6.54b}$$

with ε_{12}, ε_{11} and ε_{22} as defined above. Each of these has to be satisfied for all values of x_2 and for all times in the $x_1=0$ plane. Equation (6.54*a*) can be satisfied if the exponents in ψ_1, ψ_2 and ψ_3 are all equal at $x_1=0$. This gives

$$k_l \sin\alpha = k_l \sin\alpha' = k_t \sin\beta. \tag{6.55}$$

Therefore,

$$\alpha=\alpha' \qquad \sin\alpha/C_l=\sin\beta/C_t \tag{6.56}$$

the angle of reflection of the longitudinal wave is the same as the angle of incidence. The angle of reflection of the transverse wave, on the other hand, follows a relation similar to Snell's law for the refraction of light. By substituting (6.53) and (6.55) in (6.54*a*) one obtains

$$k_l(A_1-A_2)\sin 2\alpha - k_t A_3 \cos 2\beta=0. \tag{6.57}$$

Next we attend to equation (6.54*b*). After making replacements for $(\lambda+2\mu)$ and μ from equations (6.20) and (6.16), it becomes

$$\rho_0 C_l^2\varepsilon_{11}+\rho_0(C_l^2-2C_t^2)\varepsilon_{22}=0.$$

Now substitute for ε_{11} and ε_{22} from equations (6.50) and (6.52), and make use of the condition (6.55) which gives

$$k_l(A_1+A_2)(C_l^2-2C_t^2\sin^2\alpha)-k_t A_3 C_t^2 \sin 2\beta=0. \tag{6.58}$$

Equations (6.57) and (6.58) are useful expressions and can be solved to obtain A_2/A_1 and A_3/A_1. It is, however, straightforward to see the implications of these relations when no shear wave is reflected, i.e. $A_3=0$,

$$k_l(A_1-A_2)\sin 2\alpha=0 \tag{6.59}$$

$$k_l(A_1+A_2)(C_l^2-2C_t^2\sin^2\alpha)=0. \tag{6.60}$$

This equation can only be satisfied if (i) $\alpha=0$ or $\pi/2$, then (6.59) is satisfied automatically and (6.60) has the solution, $A_1=-A_2$. The incident longitudinal wave is reflected only as a longitudinal wave and suffers a phase change of π and (ii) the medium consists of an ideal fluid ($\mu=0$), then equivalent to (6.59) one has $\sigma_{12}=2\mu\varepsilon_{12}=0$ and (6.60) implies $A_1=-A_2$ for all angles of incidence.

Thus we notice that, except for the cases (i) and (ii), a general solution of (6.57) and (6.58) is not possible without the consideration of shear waves. These equations, after some manipulation, yield

$$\frac{A_2}{A_1}=\frac{\sin 2\alpha \sin 2\beta-(C_l/C_t)^2\cos^2 2\beta}{\sin 2\alpha\sin 2\beta+(C_l/C_t)^2\cos^2 2\beta} \tag{6.61}$$

$$\frac{A_3}{A_1}=\frac{2(C_l/C_t)\sin 2\alpha \cos 2\beta}{\sin 2\alpha \sin 2\beta+(C_l/C_t)^2 \cos^2 2\beta} \tag{6.62}$$

with

$$\frac{C_l}{C_t}=\left(\frac{2-2\nu}{1-2\nu}\right)^{1/2}.$$

A plot of the ratio of the amplitudes of the reflected and incident longitudinal waves (A_2/A_1) as a function of α for different values of ν is shown in figure 6.2. For $\nu<0.26$, there are two angles of incidence for which $A_2/A_1=0$. At these angles no longitudinal wave is reflected and all the energy of the incident longitudinal wave goes into the reflected shear waves. Conservation of energy between the incident waves and the reflected waves holds and leads to

$$\left(\frac{A_2}{A_1}\right)^2+\frac{C_t \cos \beta}{C_l \cos \alpha}\left(\frac{A_3}{A_1}\right)^2=1. \tag{6.63}$$

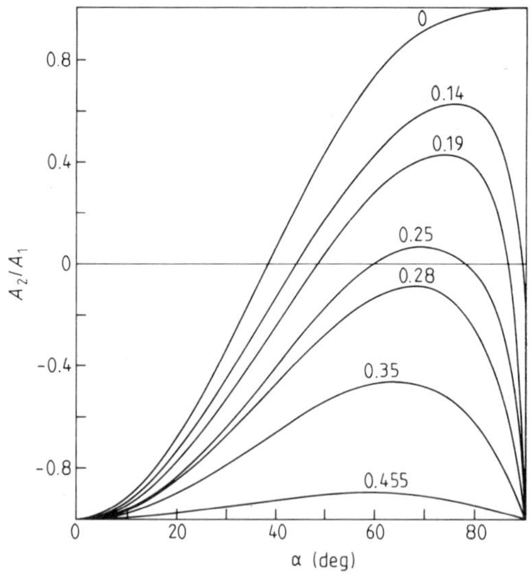

Figure 6.2 The ratio of the amplitudes (A_2/A_1) of the reflected and incident longitudinal waves as a function of the angle of incidence (α) for different values of Poisson's ratio (ν, shown on curves). Taken from Arenberg (1948).

It may be finally mentioned that the mode conversion as discussed above also occurs if a shear wave is incident upon a solid–air interface and relations equivalent to (6.57) and (6.58) can be derived. The detailed procedure is a mere repetition of what has been described here. The interested reader may enjoy working this out!

6.6.2 Reflection and refraction of waves at the plane interface between two elastic media

When a longitudinal or a shear wave is incident on a plane boundary separating two elastic media then in general due to mode conversion one gets two reflected waves and two refracted waves (shear and longitudinal) (see figure 6.3). The boundary conditions are that the normal and tangential components of the stress, and the displacements on the two sides must be equal, at $x_1 = 0$, i.e.

$$\sigma_{11}^{(1)} = \sigma_{11}^{(2)} \qquad \sigma_{12}^{(1)} = \sigma_{12}^{(2)} \qquad \sigma_{13}^{(1)} = \sigma_{13}^{(2)} \tag{6.64a}$$

$$s_i^{(1)} = s_i^{(2)} \qquad i = 1, 2, 3. \tag{6.64b}$$

The superscripts (1) and (2) refer to the two media. If α_1 and β_1 are the angles of reflection for longitudinal and shear waves, and α_2 and β_2 are the angles of refraction then analogous to (6.56) one has

$$\frac{\sin \alpha}{C_1^{(1)}} = \frac{\sin \alpha_1}{C_1^{(1)}} = \frac{\sin \beta_1}{C_t^{(1)}} = \frac{\sin \alpha_2}{C_1^{(2)}} = \frac{\sin \beta_2}{C_t^{(2)}}$$

where α is the angle of incidence and $C_1^{(1)}$, $C_1^{(2)}$, $C_t^{(1)}$ and $C_t^{(2)}$ are the velocities of the longitudinal and shear waves in medium (1) and medium (2). By performing a calculation similar to the above one can obtain relations for the amplitude of two reflected and two refracted waves in terms of the amplitude of

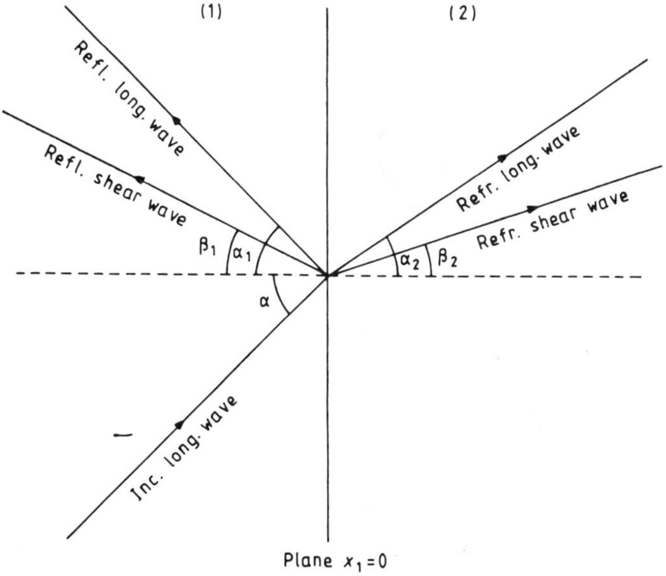

Figure 6.3 The reflection and refraction of a longitudinal wave at the interface of two elastic media.

the incident wave. For brevity, the details have been omitted here. For further discussion, the specialist books of Mason (1958) and Kolsky (1953) may be consulted.

6.6.3 Mode conversion due to scattering of elastic waves

If some inhomogeneity is created in an initially homogeneous elastic medium by including material particles of different density or of different elastic constants, it causes scattering of elastic waves. Usually the scattering becomes large with elastic anisotropy. A polycrystalline material is obviously a better scatterer than a single crystal (also called a grain) of the same material. Since different grains are differently oriented with respect to a fixed set of axes in space the values of the elastic constants differ from grain to grain. This causes scattering of elastic waves in the medium which, in general, increases when the wavelength is comparable with the grain size. The scattered waves consist of both the longitudinal and shear waves irrespective of whether the incident wave is a plane longitudinal wave or a plane shear wave. Thus, the scattering of elastic waves is also followed by mode conversion.

The scattering and the reflection of elastic waves, as discussed above, are of considerable importance. For example, infrasonic waves are used for probing the structure of the earth, and ultrasonic waves are used for testing faults in materials and in similar related problems.

Problems

6.1 Verify equations (6.23) and (6.25).

6.2 The particle displacement $\boldsymbol{s}$ in a plane irrotational wave propagating in an infinite isotropic medium is given by

$$\boldsymbol{s} = A\boldsymbol{e}\sin(kx_1 - \omega t)$$

where $\boldsymbol{e}$ is a unit vector parallel to $\boldsymbol{s}$. State what the direction of $\boldsymbol{e}$ is and obtain expressions for ε_{ij} and σ_{ij}. Express σ_{22}/σ_{11} in terms of the Poisson ratio. Does the stress σ_{ij} correspond to a uniform hydrostatic compression?

6.3 (i) Write down the equations of motion for small displacements in a cubic crystal in the absence of body forces. Assume the solution of these equations to be of the form

$$\boldsymbol{s} = s_0\boldsymbol{e}\exp\{\mathrm{i}[\omega t - k(lx_1 + mx_2 + nx_3)]\} \qquad |\boldsymbol{e}| = 1$$

where (l, m, n) are the direction cosines of $\boldsymbol{k}$. If (L, M, N) represent the direction cosine of $\boldsymbol{e}$ then write down the determinantal condition that $\boldsymbol{s}$ is the solution of the equations of motion. (The resulting equation of motion is also known as Christoffel's equation.)

(ii) Hence find the velocity ω/k of the waves propagating along the (100), (110) and (111) directions.

(iii) Show that for each of the above directions that there are two pure transverse waves and pure longitudinal waves.

6.4 Consider the reflection of plane shear waves whose direction of displacement is parallel to the plane reflecting surface at an isotropic-solid–vacuum interface. Verify that the reflected waves consist only of shear waves and that the angle of reflection is equal to the angle of incidence. What is the phase relationship between the reflected and the incident waves?

6.5 (i) A plane longitudinal wave is incident normally to a plane interface between two elastically isotropic media. For normal incidence only the longitudinal reflected and refracted waves are produced. Assuming this result derive expressions for the reflection and refraction coefficients.

(ii) How can the result derived in (i) be applied to the case where two media are liquids?

7

TRANSFORMATION LAWS FOR ELASTIC CONSTANTS AND THE EFFECT OF CRYSTAL SYMMETRY

We recall that in Chapter 4 we expressed the linear relationship between the stress and the strain tensors (Hooke's law) in the Voigt notation. The number of independent elastic constants was found to be 21 and each of these was specified with two suffixes such as c_{mn} or s_{mn}. While Voigt's notation is convenient for many purposes, it is not very helpful in discussing the way the elastic constants behave under the transformation of the coordinate axes. The usual transformation laws for tensors could not be applied to c_{mn} or s_{mn} as they are not tensors in the Voigt scheme of notation. Such knowledge, however, is very useful particularly in studying the effect of the crystal symmetry on the elastic constants.

With this purpose in mind we presently revert to tensor notation.

7.1 Elastic constants in the four-suffix notation

The generalisation of Hooke's law (see equations (4.43) and (4.44)) with σ_{ij} and ε_{ij} as tensors can be expressed as

$$\sigma_{ij}=\sum_{k,l=1}^{3} c_{ijkl}\varepsilon_{kl}\equiv c_{ijkl}\varepsilon_{kl} \tag{7.1}$$

and alternatively,

$$\varepsilon_{ij}=\sum_{k,l=1}^{3} s_{ijkl}\sigma_{kl}\equiv s_{ijkl}\sigma_{kl}. \tag{7.2}$$

As before each of equations (7.1) and (7.2) represents nine equations, each with nine terms added on the right-hand side. The coefficients c_{ijkl} and s_{ijkl} stand as usual for elastic constants and elastic compliance constants respectively. Since σ_{ij} and ε_{kl} are tensors of rank two c_{ijkl} and s_{ijkl} are tensors of rank four and hence transform according to tensor transformation rules. These in general number $3\times3\times3\times3=81$; however, this number is considerably reduced by the symmetry conditions.

For example, recalling the symmetries of the stress and strain tensor components (see §4.3),

$$\sigma_{ij}=\sigma_{ji} \qquad \text{and } \varepsilon_{ij}=\varepsilon_{ji} \tag{7.3}$$

the elastic constants can be written as

$$c_{ijkl}=c_{jikl}=c_{ijlk}=c_{jilk} \tag{7.4}$$

and similarly,

$$s_{ijkl}=s_{jikl}=s_{ijlk}=s_{jilk}. \tag{7.5}$$

From the existence of the strain-energy function (see equation (4.38))

$$\frac{\partial\sigma_{ij}}{\partial\varepsilon_{kl}}=\frac{\partial\sigma_{kl}}{\partial\varepsilon_{ij}}$$

one also infers that

$$c_{ijkl}=c_{klij} \qquad s_{ijkl}=s_{klij}. \tag{7.6}$$

Following the above equalities the numbers of independent elastic constants reduce from 81 to 21 as before. It can easily be seen that if one represents $ij\equiv m$, and $kl\equiv n$ where both m and n run from 1 to 6,

$ij;\, kl\equiv$ 11,	22, 33,	23 or 32,	31 or 13,	12 or 21	
$m, n\equiv$ 1	2 3	4	5	6	

then the two-suffix notation follows:

$$c_{ijkl}=c_{mn} \tag{7.7a}$$

$$s_{ijkl}=s_{mn} \quad \text{when both } m \text{ and } n=1, 2, 3 \tag{7.7b}$$

$$=\tfrac{1}{2}s_{mn} \quad \text{when either } m \text{ or } n=4, 5, 6 \tag{7.7c}$$

$$=\tfrac{1}{4}s_{mn} \quad \text{when both } m \text{ and } n=4, 5, 6. \tag{7.7d}$$

Bearing in mind the requirements, one uses either the four-suffix or the two-suffix notation. All the 21 independent elastic constants, both in four-suffix and two-suffix notation, are given below:

$$\begin{array}{llllll}
c_{1111} & c_{1122} & c_{1133} & c_{1123} & c_{1113} & c_{1112} \\
c_{11} & c_{12} & c_{13} & c_{14} & c_{15} & c_{16} \\
 & c_{2222} & c_{2233} & c_{2223} & c_{2213} & c_{2212} \\
 & c_{22} & c_{23} & c_{24} & c_{25} & c_{26} \\
 & & c_{3333} & c_{3323} & c_{3313} & c_{3312} \\
 & & c_{33} & c_{34} & c_{35} & c_{36} \\
 & & & c_{2323} & c_{2313} & c_{2312} \\
 & & & c_{44} & c_{45} & c_{46} \\
 & & & & c_{1313} & c_{1312} \\
 & & & & c_{55} & c_{56} \\
 & & & & & c_{12}c_{12} \\
 & & & & & c_{66}.
\end{array} \tag{7.8}$$

The matrix (7.8) is diagonally symmetrical.

7.2 Transformation laws for elastic constants

Equations (7.1) and (7.2) are proper tensor relations where c_{ijkl} and s_{ijkl} represent components of tensors of rank four. Referring equation (7.1) to the primed coordinate system one may write

$$\sigma'_{ij} = c'_{ijkl}\varepsilon'_{kl}. \tag{7.9}$$

Since σ_{ij} are components of a second-rank tensor, it follows that

$$\begin{aligned}\sigma'_{ij} &= a_{im}a_{jn}\sigma_{mn} \\ &= a_{im}a_{jn}c_{mnop}\varepsilon_{op} \\ &= a_{im}a_{jn}c_{mnop}a_{ko}a_{lp}\varepsilon'_{kl}\end{aligned} \tag{7.10}$$

where all the i, j, k, l, m, n, o and p vary from 1 to 3. From equations (7.9) and (7.10),

$$c'_{ijkl} = a_{im}a_{jn}a_{ko}a_{lp}c_{mnop}. \tag{7.11}$$

Similarly, starting with equation (7.2) one obtains

$$s'_{ijkl} = a_{im}a_{jn}a_{ko}a_{lp}s_{mnop}. \tag{7.12}$$

In the above equations the a's as usual stand for the direction cosines. Thus c_{ijkl} and s_{ijkl} transform like a tensor of rank four.

7.3 Reduction of independent elastic constants because of crystal symmetry

In §3.7 we discussed how the components of a second-rank tensor (thermal-expansion coefficient as an example) reduce depending upon the symmetry of the crystal. Some useful references dealing with similar problems are also given there in detail. Here we briefly address the problem of the reduction of the elastic constants conforming to crystal symmetry. There are, in general, seven crystal systems namely the triclinic, monoclinic, orthorhombic, tetragonal, trigonal, hexagonal and cubic systems. We shall, however, restrict ourselves to a few crystal symmetries which occur frequently in nature.

7.3.1 Monoclinic crystal

As pointed out earlier a monoclinic crystal has an axis of symmetry for rotations of 180°, say this is the x_3-axis. The crystal properties thus remain invariant for the transformations: $x'_1 = -x_1$, $x'_2 = -x_2$ and $x'_3 = x_3$. The various direction cosines are (see equation (2.10))

$$\begin{gathered} a_{11} = -1 \qquad a_{22} = -1 \qquad a_{33} = 1 \\ a_{12} = a_{21} = a_{13} = a_{31} = a_{23} = a_{32} = 0. \end{gathered} \tag{7.13}$$

In view of (7.13) the transformation matrix (7.11) ca now be written as

$$c'_{ijkl} = a_{im}a_{jn}a_{ko}a_{lp}c_{mnop}$$
$$= a_{ii}a_{jj}a_{kk}a_{ll}c_{ijkl}$$

or

$$c'_{ijkl} = \mp c_{ijkl}. \qquad (7.14)$$

The negative sign corresponds to that when the suffix 3 occurs an odd number of times among *ijkl*. Fulfilling the symmetry requirements ($c'_{ijkl} = c_{ijkl}$) it is evident that the c's having an odd number of the suffixes 3 must be zero. More explicitly if $ijkl = 1113$; 1123; 2213; 2223; 3313; 3323; 2312; and 1312 then correspondingly one has $c_{15} = c_{14} = c_{25} = c_{24} = c_{35} = c_{34} = c_{46} = c_{56} = 0$. The non-vanishing c's for the monoclinic crystal are, therefore,

$$\begin{matrix} c_{11} & c_{12} & c_{13} & 0 & 0 & c_{16} \\ & c_{22} & c_{23} & 0 & 0 & c_{26} \\ & & c_{33} & 0 & 0 & c_{36} \\ & & & c_{44} & c_{45} & 0 \\ & & & & c_{55} & 0 \\ & & & & & c_{66} \end{matrix} . \qquad (7.15)$$

The matrix (7.15) is diagonally symmetrical. There are in all 13 independent elastic constants for monoclinic crystals.

7.3.2 Orthorhombic crystal

Orthorhombic crystals possess three perpendicular axes of symmetry for the 180° rotation. Let us choose the x_1-, x_2- and x_3-axes as the axes of symmetry. Taking the x_3-axis as the axis of symmetry the resulting matrix for c_{ijkl} is given in (7.15). Similarly if one considers a rotation of 180° about the x_2-axis then in (7.15) all those c_{ijkl} will vanish for which the suffix 2 occurs an odd number of times,

$$ijkl = 11\ 12,\quad 22\ 12,\quad 33\ 12,\ 23\ 13$$
$$c_{mn} = c_{16} = c_{26} = c_{36} = c_{45} = 0. \qquad (7.16)$$

No more elastic constants, however, vanish by virtue of the third axis of symmetry. Thus taking (7.15) and (7.16) together, there are only nine independent elastic constants for the orthorhombic crystal. The resulting matrix for c_{mn} is

$$\begin{matrix} c_{11} & c_{12} & c_{13} & 0 & 0 & 0 \\ & c_{22} & c_{23} & 0 & 0 & 0 \\ & & c_{33} & 0 & 0 & 0 \\ & & & c_{44} & 0 & 0 \\ & & & & c_{55} & 0 \\ & & & & & c_{66} \end{matrix}. \tag{7.17}$$

7.3.3 *Cubic crystal*

A cubic crystal is endowed with three mutually perpendicular axes of symmetry for 90° rotations. The number of independent elastic constants can be obtained here almost by inspection. For convenience let us choose the axes x_1, x_2, x_3 parallel to the cube edges. These axes are now explicitly the axes of symmetry for a 180° rotation. Thus, the non-vanishing elastic constants for cubic crystals are the same as for orthorhombic crystals.

Furthermore, cubic symmetry implies that three cube edges and hence the axes x_1, x_2 and x_3 are equivalent. It does not matter which axis is labelled as 1 or 2 or 3. Hence,

$$\begin{aligned} c_{1111} &= c_{2222} = c_{3333} \\ c_{1122} &= c_{1133} = c_{2233} \\ c_{2323} &= c_{2121} = c_{1313}. \end{aligned} \tag{7.18}$$

The matrix for c_{mn} is thus,

$$\begin{matrix} c_{11} & c_{12} & c_{12} & 0 & 0 & 0 \\ & c_{11} & c_{12} & 0 & 0 & 0 \\ & & c_{11} & 0 & 0 & 0 \\ & & & c_{44} & 0 & 0 \\ & & & & c_{44} & 0 \\ & & & & & c_{44} \end{matrix}. \tag{7.19}$$

There are just three independent elastic constants (c_{11}, c_{12} and c_{44}) for cubic crystals.

7.3.4 *Hexagonal crystal*

A hexagonal crystal has an axis of symmetry for rotations of 60°. The direction cosines are (see equation (3.100))

$$\begin{gathered} a_{11}=a_{22}=\tfrac{1}{2} \qquad a_{12}=-a_{21}=\sqrt{3}/2 \qquad a_{33}=1 \\ a_{13}=a_{23}=a_{31}=a_{32}=0. \end{gathered} \tag{7.20}$$

When these coefficients are substituted in equation (7.11) one obtains the non-vanishing components of c_{ijkl}. Omitting the mathematical details we furnish here only the matrix c_{mn} for hexagonal crystals:

$$\begin{matrix} c_{11} & c_{12} & c_{13} & 0 & 0 & 0 \\ & c_{11} & c_{13} & 0 & 0 & 0 \\ & & c_{33} & 0 & 0 & 0 \\ & & & c_{44} & 0 & 0 \\ & & & & c_{44} & 0 \\ & & & & & \tfrac{1}{2}(c_{11}-c_{12}) \end{matrix} . \tag{7.21}$$

Finally, for completeness, the elastic constants for some of the crystals are given in table 7.1.

7.4 Condition for elastic isotropy of a crystal

In an isotropic crystal all the directions are equivalent and hence the elastic properties are the same in every direction. Obviously, the elastic-constants matrix must be at least that of a cubic crystal. Consider the transformation

Table 7.1 Elastic constants† of crystals (in units of $10^{11}\ \mathrm{dyn\,cm^{-2}}$) at room temperature.

Cubic system	c_{11}	c_{12}	c_{44}		
Potassium	0.41	0.33	0.26		
Lead	4.76	4.03	1.44		
Sodium chloride	4.64	1.26	1.27		
Aluminium	10.82	6.21	2.84		
Copper	17.02	11.49	6.09		
Iron	22.81	13.35	11.08		
Diamond	94.90	15.10	52.10		
Hexagonal system	c_{11}	c_{12}	c_{13}	c_{33}	c_{55}
Ice (257 K)	1.38	0.71	0.58	1.49	0.32
Beryllium	30.80	5.80	8.70	35.70	11.00

† The source of the data is the compilation by Simons and Wang (1971).

equation (7.11)

$$c'_{ijkl} = a_{im}a_{jn}a_{ko}a_{lp}c_{mnop}.$$

For $i=j=k=l=1$, and employing the equalities of equation (7.18) one obtains

$$\begin{aligned} c'_{1111} = {} & c_{1111}(a_{11}^4 + a_{12}^4 + a_{13}^4) \\ & + c_{1122}(2a_{11}^2 a_{12}^2 + 2a_{11}^2 a_{13}^2 + 2a_{12}^2 a_{13}^2) \\ & + c_{2323}(4a_{12}^2 a_{13}^2 + 4a_{11}^2 a_{13}^2 + 4a_{11}^2 a_{12}^2). \end{aligned} \tag{7.22}$$

If we denote the direction cosines, $a_{11} \equiv l$, $a_{12} \equiv m$, $a_{13} \equiv n$ and recall the identities

$$l^2 + m^2 + n^2 = 1 \tag{7.23a}$$

$$(l^2 + m^2 + n^2)^2 = l^4 + m^4 + n^4 + 2(l^2m^2 + m^2n^2 + n^2l^2) = 1 \tag{7.23b}$$

then equation (7.22) becomes

$$\begin{aligned} c'_{1111} &= c_{1111}(l^4 + m^4 + n^4) + 2(c_{1122} + 2c_{2323})(l^2m^2 + m^2n^2 + n^2l^2) \\ &= c_{1111} - 2(c_{1111} - c_{1122} - 2c_{2323})(l^2m^2 + m^2n^2 + n^2l^2) \end{aligned} \tag{7.24}$$

or in the two-suffix notation,

$$c'_{11} = c_{11} - 2(c_{11} - c_{12} - 2c_{44})(l^2m^2 + m^2n^2 + n^2l^2). \tag{7.25}$$

However, for the crystal to be elastically isotropic one must have

$$c'_{11} = c_{11} \tag{7.26}$$

which is possible if

$$c_{11} - c_{12} - 2c_{44} = 0$$

the same as in equation (4.70). This is known as the elastic isotropic condition. Sometimes it is equivalently written as

$$\gamma = 2c_{44}/(c_{11} - c_{12}) \tag{7.27}$$

where γ is called the anisotropic factor. It takes the value of one for isotropic materials. The values of γ for selected materials are given in table 7.2. It may be noted that most of the actual single crystals are far from isotropy. Only for tungsten is γ very close to unity.

7.5 Averaging of single-crystal elastic constants

Most materials which are found to occur naturally are polycrystalline. This means that they consist of a large number of small single crystals (or grains) oriented more or less randomly. The material then behaves, to a first-order approximation, as a macroscopically isotropic body. We have already seen

Table 7.2 Values† of the anisotropic factor.

System	γ
Potassium	6.5
Lead	3.9
Iron	2.3
Copper	2.2
Diamond	1.3
Aluminium	1.2
Tungsten	1.0
Sodium chloride	0.7

† Obtained from the elastic-constants data. The source is Simons and Wang (1971).

(§7.3) that the number of single-crystal elastic constants are many and depend upon the crystal symmetry. It is therefore desirable to calculate the average elastic constants; one for longitudinal motion, $\langle c_{11}\rangle_{\text{av}}$, and one for shear motion, $\langle c_{44}\rangle_{\text{av}}$, for a given crystal symmetry. These are, in turn, useful for estimating many bulk elastic properties of materials as well as in the study of the attenuation of elastic waves.

To obtain $\langle c_{11}\rangle_{\text{av}}$ and $\langle c_{44}\rangle_{\text{av}}$ we shall use the general tensor relation (7.11) where the averaging has to be performed over all possible orientations of the grains. This could be achieved by taking the direction cosines for rotations in terms of Euler's angles, α $(0\leqslant\alpha\leqslant 2\pi)$, β $(0\leqslant\beta\leqslant\pi)$ and γ $(0\leqslant\gamma\leqslant 2\pi)$, see problem (P2.2). A general expression for the averaged elastic constants may, therefore, be written as,

$$\langle c'_{ijkl}\rangle_{\text{av}}=\int_0^{2\pi}\frac{\mathrm{d}\gamma}{2\pi}\int_0^{2\pi}\frac{\mathrm{d}\alpha}{2\pi}\int_0^{\pi}\frac{\sin\beta\,\mathrm{d}\beta}{2}\,(c'_{ijkl}). \tag{7.28}$$

To make proper use of equation (7.28), c'_{ijkl} on the right-hand side should be expanded in terms of the a's and c's through equation (7.11). Then inserting the direction cosine a's from equation (P2.2) the integration is carried out. For grains of orthorhombic symmetry one, for instance, obtains

$$\begin{aligned}\langle c_{11}\rangle_{\text{av}}&=\langle c'_{1111}\rangle\\&=\tfrac{1}{3}(c_{11}+c_{22}+c_{33})-\tfrac{2}{5}a\end{aligned} \tag{7.29}$$

$$\begin{aligned}\langle c_{44}\rangle_{\text{av}}&=\langle c'_{2323}\rangle\\&=\tfrac{1}{3}(c_{44}+c_{55}+c_{66})+\tfrac{1}{5}a\end{aligned} \tag{7.30}$$

where

$$3a = c_{11} + c_{22} + c_{33} - (c_{12} + c_{13} + c_{23}) - 2(c_{44} + c_{55} + c_{66}). \qquad (7.31)$$

Equations (7.29) to (7.31) can also be used to obtain average values for cubic and hexagonal crystals. One simply needs to substitute various identities of c's from (7.19) and (7.21) respectively.

For a comprehensive review, the reader is referred to Chapter 15 of Musgrave (1970) and other references given there.

Problems

7.1 Verify equation (7.7).

7.2 Bearing in mind that c_{ijkl} are components of a fourth-rank tensor show that under the orthogonal transformation of coordinate axes,

$$\begin{aligned} c'_{11} &= c_{11}(a_{11}^2 + a_{12}^2)^2 + c_{33}a_{13}^4 \\ &\quad + (2c_{13} + 4c_{44})(a_{11}^2 a_{13}^2 + a_{12}^2 a_{13}^2) \end{aligned} \qquad (\text{P7.1})$$

$$\begin{aligned} c'_{44} &= c_{11}[(a_{11}a_{21} + a_{12}a_{22})^2 + \tfrac{1}{2}(a_{11}a_{22} - a_{12}a_{21})^2] \\ &\quad - \tfrac{1}{2}c_{12}(a_{11}a_{22} - a_{12}a_{21})^2 \\ &\quad + 2c_{13}(a_{11}a_{13}a_{21}a_{23} + a_{12}a_{13}a_{22}a_{23}) + c_{33}a_{13}^2 a_{23}^2 \\ &\quad + c_{44}[(a_{12}a_{23} + a_{13}a_{22})^2 + (a_{13}a_{21} + a_{11}a_{23})^2] \end{aligned} \qquad (\text{P7.2})$$

for a crystal having hexagonal symmetry. c_{mn} are elastic constants in the Voigt notation.

7.3 Show that the reciprocal of Young's modulus for a cubic crystal in a direction whose direction cosines are l, m, n referred to the cube axes as coordinate axes is given by

$$E^{-1} = s_{11} - 2(s_{11} - s_{12} - \tfrac{1}{2}s_{44})(l^2m^2 + m^2n^2 + n^2l^2). \qquad (\text{P7.3})$$

8

THE THERMODYNAMICS OF SOLIDS

The basic laws of thermodynamics[†] which are well known to us for fluids apply equally well to solids. The thermodynamic state of a fluid is described by any two of the three state variables: P (pressure), V (volume) and T (temperature). These variables are not all independent, rather they are related by virtue of the equation of state,

$$f(P, V, T) = 0. \tag{8.1}$$

For solids, however, these variables are replaced by several other variables. For instance, the deformation in solids is described by six strain components and, equivalent to hydrostatic pressure, there are six stress components. In effect, one deals with seven independent variables: six ε_m and T or six σ_m and T to describe a thermodynamic state of a solid body. Although the extension of the thermodynamic formulae to solids is straightforward nevertheless it is instructive to derive them. In the sections to follow the basic laws of thermodynamics in terms of ε_m and σ_m will be developed with an application to thermoelastic damping. For more applications the reader may refer to work such as McLellan (1980) and Wallace (1972).

8.1 Mathematical forms of first and second laws of thermodynamics

The first law of thermodynamics simply advocates the energy conservation law. For any quasi-static process[‡] one writes

$$\mathrm{d}Q = \mathrm{d}U + \mathrm{d}W \tag{8.2}$$

where $\mathrm{d}Q$ is the amount of heat entering into the system, $\mathrm{d}U$ represents the change occurring in the internal energy and $\mathrm{d}W$ is the work done by the system.

The second law, on the other hand, implies the existence of a characteristic function S (called the entropy) such that in an infinitesimal reversible process

$$\mathrm{d}Q = T\,\mathrm{d}S \tag{8.3}$$

[†] Readers may familiarise themselves with various thermodynamic quantities from a specialised book such as Zemansky (1968).

[‡] In a quasi-static process the system is in thermodynamic equilibrium at every instant of time.

where T is the absolute temperature. Equations (8.2) and (8.3) may be fused together to make a combined statement of the first and second laws of thermodynamics,

$$T\,\mathrm{d}S = \mathrm{d}U + \mathrm{d}W. \tag{8.4}$$

For a hydrostatic system $\mathrm{d}W = P\,\mathrm{d}V$ and therefore equation (8.4) becomes

$$T\,\mathrm{d}S = \mathrm{d}U + P\,\mathrm{d}V. \tag{8.5}$$

However, for a solid $\mathrm{d}W$ is defined differently. We recall that if the strains ε_m were changed to $\varepsilon_m + \mathrm{d}\varepsilon_m$ ($m = 1, 2, \ldots, 6$), the amount of work done on the system per unit volume is (see equation (4.41))

$$\mathrm{d}W = \sum_{m=1}^{6} \sigma_m\,\mathrm{d}\varepsilon_m$$

where σ_m, as usual, are stress components in the Voigt notation. Obviously, the work done by the solid is equal to $-\mathrm{d}W$, and hence equation (8.4) becomes

$$T\,\mathrm{d}S = \mathrm{d}U - \sum_{m} \sigma_m\,\mathrm{d}\varepsilon_m. \tag{8.6}$$

This is a basic thermodynamic relation for solids. It may be emphasised that in a solid it is customary to refer all extensive variables to the amount of material whose volume before deformation was unity. This is called the reference state. Therefore, S and U in equation (8.5) refer to the amount of material contained in unit volume before deformation.

8.2 Some useful thermodynamic relations

8.2.1 Definition of adiabatic and isothermal elastic constants

In fluids one defines the bulk modulus B either at constant temperature T (known as isothermal), or at constant entropy S (known as adiabatic), i.e.

$$B_T = -V(\partial P/\partial V)_T \qquad B_S = -V(\partial P/\partial V)_S. \tag{8.7}$$

By analogy we define the isothermal and adiabatic elastic constants for solids as

$$c_{mn}^{T} = (\partial\sigma_m/\partial\varepsilon_n)_T \qquad c_{mn}^{S} = (\partial\sigma_m/\partial\varepsilon_n)_S. \tag{8.8}$$

The isothermal and adiabatic compliance constants are, similarly, defined as

$$s_{mn}^{T} = (\partial\varepsilon_m/\partial\sigma_n)_T \qquad s_{mn}^{S} = (\partial\varepsilon_m/\partial\sigma_n)_S \tag{8.9}$$

where the variables to be kept constant during the differentiation are obvious. In the first expression of (8.8), the differential has to be obtained with respect to

ε_n keeping T and the remaining five ε_n constant and similarly for the other expressions.

As we proceed further we shall obtain some useful thermodynamic relations for solids.

8.2.2 Helmholtz and Gibbs free-energy functions

We shall recall an important thermodynamic function, the Helmholtz free energy A,

$$A = U - TS \tag{8.10}$$

or

$$\mathrm{d}A = \mathrm{d}U - T\,\mathrm{d}S - S\,\mathrm{d}T. \tag{8.11}$$

Making a substitution for $\mathrm{d}U - T\,\mathrm{d}S$ from (8.6), one has

$$\mathrm{d}A = \sum_m \sigma_m\,\mathrm{d}\varepsilon_m - S\,\mathrm{d}T. \tag{8.12}$$

Thus, A is a function of seven independent variables, temperature and six strain components ε_m,

$$A = A(T, \varepsilon_m; m = 1, 2, \ldots, 6). \tag{8.13}$$

Another important thermodynamic function is the Gibbs free energy G,

$$G = A - \sum_m \sigma_m \varepsilon_m \tag{8.14}$$

or

$$\mathrm{d}G = \mathrm{d}A - \sum_m \sigma_m\,\mathrm{d}\varepsilon_m - \sum_m \varepsilon_m\,\mathrm{d}\sigma_m \tag{8.15}$$

which on using equation (8.12) reduces to

$$\mathrm{d}G = -S\,\mathrm{d}T - \sum_m \varepsilon_m\,\mathrm{d}\sigma_m. \tag{8.16}$$

Thus one writes

$$G = G(T, \sigma_m; m = 1, 2, \ldots, 6). \tag{8.17}$$

Now from equations (8.12) and (8.16) we have respectively

$$S = -(\partial A/\partial T)_\varepsilon \qquad \sigma_m = (\partial A/\partial \varepsilon_m)_T \tag{8.18a}$$

and

$$S = -(\partial G/\partial T)_\sigma \qquad \varepsilon_m = -(\partial G/\partial \sigma_m)_T \tag{8.18b}$$

where the subscripts ε and σ mean that all the strain and stress components are to be kept constant.

Since $\mathrm{d}A$ is a perfect differential, we therefore further write

$$-(\partial\sigma_m/\partial T)_\varepsilon = (\partial S/\partial\varepsilon_m)_T \equiv \mathscr{F}_m \tag{8.19}$$

where $\mathscr{F}_m$ may be regarded as the coefficients of thermal stress. Like σ_m, $\mathscr{F}_m$ are also components of a tensor of rank two. Similarly, from equation (8.16) one has

$$(\partial\varepsilon_m/\partial T)_\sigma = (\partial S/\partial\sigma_m)_T \equiv \alpha_m. \tag{8.20}$$

Then α_m are known as the coefficients of thermal strain and, like the strain tensor, α_m are also components of a second-rank tensor.

8.2.3 Relation between the adiabatic and the isothermal elastic constants

Consider T and ε as independent variables and $S(T, \varepsilon)$ and $\sigma(T, \varepsilon)$ as dependent variables. The perfect differential of S can be written as

$$\begin{aligned}\mathrm{d}S &= \left(\frac{\partial S}{\partial T}\right)_\varepsilon \mathrm{d}T + \sum_n \left(\frac{\partial S}{\partial\varepsilon_n}\right)_T \mathrm{d}\varepsilon_n \\ &= \left(\frac{C_\varepsilon}{T}\right)\mathrm{d}T + \sum_n \mathscr{F}_n\,\mathrm{d}\varepsilon_n \end{aligned} \tag{8.21}$$

with

$$C_\varepsilon = T(\partial S/\partial T)_\varepsilon \qquad \mathscr{F}_n = (\partial S/\partial\varepsilon_n)_T. \tag{8.22}$$

C_ε represents the specific heat at constant strain. It is analogous to C_V $(= T(\partial S/\partial T)_V)$ for fluids.

Similarly, the perfect differential of σ_m can be written as

$$\begin{aligned}\mathrm{d}\sigma_m &= \sum_n \left(\frac{\partial\sigma_m}{\partial\varepsilon_n}\right)_T \mathrm{d}\varepsilon_n + \left(\frac{\partial\sigma_m}{\partial T}\right)_\varepsilon \mathrm{d}T \\ &= \sum_n c^T_{mn}\,\mathrm{d}\varepsilon_n - \mathscr{F}_m\,\mathrm{d}T. \end{aligned} \tag{8.23}$$

Eliminating $\mathrm{d}T$ from equations (8.21) and (8.23), one has

$$\mathrm{d}\sigma_m = \sum_n c^T_{mn}\,\mathrm{d}\varepsilon_n - \frac{\mathscr{F}_m T}{C_\varepsilon}\left(\mathrm{d}S - \sum_n \mathscr{F}_n\,\mathrm{d}\varepsilon_n\right). \tag{8.24}$$

This is now used in equation (8.8) to obtain an expression for the difference between the adiabatic and the isothermal elastic constants,

$$c^S_{mn} - c^T_{mn} = \mathscr{F}_m\mathscr{F}_n T/C_\varepsilon. \tag{8.25}$$

In addition, if we divide equation (8.23) for each m by $\mathrm{d}T$ and keep all the stress components, σ, constant then since $(\partial \varepsilon_n/\partial T)_\sigma = \alpha_n$, one has

$$\mathscr{F}_m = \sum_n c^T_{mn}\alpha_n. \tag{8.26}$$

Next considering σ and T as independent variables one obtains equations similar to (8.21) and (8.23):

$$\mathrm{d}S = \left(\frac{C_\sigma}{T}\right)\mathrm{d}T + \sum_m \alpha_m \,\mathrm{d}\sigma_m \tag{8.27}$$

$$\mathrm{d}\varepsilon_m = \sum_n s^T_{mn}\,\mathrm{d}\sigma_n + \alpha_m\,\mathrm{d}T \tag{8.28}$$

where

$$C_\sigma = T(\partial S/\partial T)_\sigma \tag{8.29}$$

is the specific heat of solids at constant stress. C_σ is analogous to C_P $(=T(\partial S/\partial T)_P)$ for fluids. Equations (8.27) and (8.28) further yield

$$s^S_{mn} - s^T_{mn} = -\alpha_m\alpha_n T/C_\sigma \tag{8.30}$$

and

$$\alpha_m = \sum_n s^T_{mn}\mathscr{F}_n. \tag{8.31}$$

8.2.4 Relation between the specific heats at constant stress and constant strain

Substitution of $\mathrm{d}\sigma_m$ from (8.23) into equation (8.27) allows one to write

$$\mathrm{d}S = \left(\frac{C_\sigma}{T}\right)\mathrm{d}T + \sum_m \alpha_m\left(\sum_n c^T_{mn}\,\mathrm{d}\varepsilon_n - \mathscr{F}_m\,\mathrm{d}T\right) \tag{8.32}$$

or,

$$\left(\frac{\partial S}{\partial T}\right)_\varepsilon = \frac{C_\sigma}{T} - \sum_m \alpha_m\mathscr{F}_m. \tag{8.33}$$

Bearing in mind that $T(\partial S/\partial T)_\varepsilon = C_\varepsilon$, the above equation readily gives the difference between the specific heats at constant stress and constant strain,

$$\begin{aligned} C_\sigma - C_\varepsilon &= T\sum_m \alpha_m\mathscr{F}_m \\ &= T\sum_{m,n}\alpha_m\alpha_n c^T_{mn}. \end{aligned} \tag{8.34}$$

The various thermodynamic relations derived above are valid for crystals of all symmetries. A considerable simplification, however, occurs depending upon the symmetry of the crystal. In the section to follow we consider cubic and isotropic materials as examples.

8.3 Thermodynamic relations for cubic and isotropic solids

We recall from §3.7 that for a cubic or for an isotropic material there is only one independent component of the thermal-expansion coefficients,

$$\begin{aligned} \alpha_1 &= \alpha_2 = \alpha_3 = \bar{\alpha} \text{ (say)} \\ \alpha_4 &= \alpha_5 = \alpha_6 = 0. \end{aligned} \tag{8.35}$$

Using this with equation (8.26) one writes

$$\begin{aligned} \mathscr{F}_1 &= \mathscr{F}_2 = \mathscr{F}_3 = \mathscr{F} \text{ (say)} \\ \mathscr{F}_4 &= \mathscr{F}_5 = \mathscr{F}_6 = 0. \end{aligned} \tag{8.36}$$

Now consider equation (8.21),

$$\begin{aligned} dS &= \frac{C_\varepsilon}{T}\, dT + \sum_n \mathscr{F}_n\, d\varepsilon_n \\ &= (C_\varepsilon/T)\, dT + \mathscr{F}(d\varepsilon_1 + d\varepsilon_2 + d\varepsilon_3) \\ &= (C_\varepsilon/T)\, dT + \mathscr{F}\, d\Delta \end{aligned} \tag{8.37}$$

where $\Delta = \varepsilon_1 + \varepsilon_2 + \varepsilon_3$ is as usual the dilatation. Equation (8.37) can be written as

$$T(\partial S/\partial T)_V = C_\varepsilon$$

or

$$C_V = C_\varepsilon. \tag{8.38}$$

Similarly equations (8.27) and (8.36) yield,

$$C_P = C_\sigma. \tag{8.39}$$

Thus for isotropic and cubic materials,

$$C_P - C_V = C_\sigma - C_\varepsilon. \tag{8.40}$$

Recalling equation (8.34) at this stage allows us to write,

$$\begin{aligned} C_P - C_V &= T \sum_{m,n=1}^{6} \alpha_m \alpha_n c_{mn}^T \\ &= T\bar{\alpha}^2 \sum_{m,n=1}^{3} c_{mn}^T \\ &= 9T\bar{\alpha}^2 \left(\frac{c_{11} + 2c_{12}}{3}\right)_T. \end{aligned} \tag{8.41}$$

The bracketed term on the right-hand side of (8.41) is the bulk modulus, B_T. Since $\bar{\alpha}=\frac{1}{3}\alpha$ (α is the coefficient of volume expansion), equation (8.41) reduces to

$$C_P - C_V = T\alpha^2 B_T \tag{8.42}$$

which is the required relation for the difference between specific heat at constant pressure and constant volume for cubic or isotropic materials. Relation (8.42) may be noted to be the same as for fluids.

Furthermore, equations (8.36) and (8.25) can be employed to obtain the difference between the adiabatic and isothermal elastic constants,

$$c_{mn}^S - c_{mn}^T = (T/C_\varepsilon)\mathscr{F}_m\mathscr{F}_n = (T/C_V)\mathscr{F}^2. \tag{8.43}$$

Making a substitution for $\mathscr{F}$ from (8.26),

$$\mathscr{F} = \bar{\alpha}\sum_{n=1}^{3} c_{1n}^T = \bar{\alpha}(c_{11}+2c_{12}) = 3\bar{\alpha}B_T = \alpha B_T$$

we obtain,

$$c_{mn}^S - c_{mn}^T = (T/C_V)\alpha^2 B_T^2 \tag{8.44}$$

the relation we wished to derive. For $m, n = 1, 2, 3$, it can be rewritten as

$$c_{11}^S - c_{11}^T = c_{12}^S - c_{12}^T = B_S - B_T = (T/C_V)\alpha^2 B_T^2. \tag{8.45}$$

For $m, n = 4, 5, 6$; $\mathscr{F}_4 = \mathscr{F}_5 = \mathscr{F}_6 = 0$ and therefore

$$c_{44}^S - c_{44}^T = 0. \tag{8.46}$$

It is thus evident that there is no difference between the adiabatic and the isothermal shear modulus for cubic or isotropic materials.

Finally, from equations (8.41) and (8.45) we obtain

$$B_S - B_T = [(C_P - C_V)/C_V]B_T = (\gamma - 1)B_T \tag{8.47a}$$

or

$$B_S/B_T = C_P/C_V = \gamma \tag{8.47b}$$

a result similar to fluids. However, owing to different values of γ for a fluid and for a solid, it provides a differential insight. For example, $\gamma_{\text{air}} \simeq 1.4$, i.e. the adiabatic bulk modulus (B_S) of air is about 40% higher than the isothermal bulk modulus (B_T); for liquids $\gamma_{\text{liq}} \simeq 1.1$, i.e. B_S is only higher than B_T by about 10%.

For solids, on the other hand, $\gamma - 1 \simeq 10^{-2}$ at room temperature and decreases more with decreasing temperature. Equation (8.47a) therefore reduces to

$$B_S \simeq B_T \qquad T \leqslant \text{room temperature.} \tag{8.48}$$

We may infer that the values of the adiabatic and isothermal elastic constants for solids are approximately the same. This seems to suggest that the calculation of the velocity of propagation of a wave in a solid either assuming the vibrations to be purely adiabatic or purely isothermal does not involve appreciable error. In practice, however, it is not true because any vibratory motion such as the vibrations of a string, of a plate, of a drum, etc are neither purely adiabatic nor purely isothermal. The part of the material which undergoes compression experiences a rise in temperature and heat flows from compressed to expanded regions in an attempt to equalise the temperature. During the process some part of the mechanical energy is lost irreversibly into heat energy leading to the damping of vibrations and hence wave motion. The damping of mechanical vibrations due to heat flow is called thermoelastic damping or thermoelastic relaxation. A review of the problem of damping for various types of vibrations can be found in books like Mason (1958) and Bhatia (1967). We shall, however, provide a brief discussion of the thermoelastic damping of plane waves in the following section.

8.4 Attenuation of plane waves due to thermal conductivity

As we have just mentioned the propagation of a wave accompanied with compression and extension of the medium suffers a loss in energy. The compressed part becomes hot which leads to a flow of heat into the extended region. As a consequence of thermal conduction energy is lost which results in the damping of vibrations. We shall see shortly that the shear wave which does not involve a change in volume incurs no loss due to heat flow.

We start with two thermodynamic relations which we have already deduced earlier (see equations (8.21) and (8.23))

$$\mathrm{d}S = \left(\frac{C_\varepsilon}{T}\right)\mathrm{d}T + \sum_n \mathscr{F}_n\, \mathrm{d}\varepsilon_n \qquad m, n = 1, 2, \ldots, 6$$

$$\mathrm{d}\sigma_m = \sum_n c^T_{mn}\, \mathrm{d}\varepsilon_n - \mathscr{F}_m\, \mathrm{d}T$$

For simplicity, we specialise to an elastically isotropic material for which $c_{11} = c_{12} + 2c_{44}$: $\mathscr{F}_m$ is defined in equation (8.36); and $C_\varepsilon = C_V$, therefore

$$\mathrm{d}S = \left(\frac{C_V}{T}\right)\mathrm{d}T + \mathscr{F}\, \mathrm{d}\Delta \tag{8.49a}$$

$$\mathrm{d}\sigma_m = c^T_{12}\, \mathrm{d}\Delta + 2c^T_{44}\, \mathrm{d}\varepsilon_m - \mathscr{F}\, \mathrm{d}T \qquad m = 1, 2, 3 \tag{8.49b}$$

$$\mathrm{d}\sigma_m = c^T_{44}\, \mathrm{d}\varepsilon_n \qquad m = 4, 5, 6 \tag{8.49c}$$

where $\Delta = \varepsilon_1 + \varepsilon_2 + \varepsilon_3$ is, as before, the dilatation. Let us recall the equation of motion (see equation (6.12)) which in the absence of body forces ($F_i = 0$) may be written as (in the Voigt notation)

$$\rho_0 \frac{\partial^2 s_1}{\partial t^2} = \frac{\partial \sigma_1}{\partial x_1} + \frac{\partial \sigma_6}{\partial x_2} + \frac{\partial \sigma_5}{\partial x_3} \tag{8.50a}$$

$$\rho_0 \frac{\partial^2 s_2}{\partial t^2} = \frac{\partial \sigma_6}{\partial x_1} + \frac{\partial \sigma_2}{\partial x_2} + \frac{\partial \sigma_4}{\partial x_3} \tag{8.50b}$$

$$\rho_0 \frac{\partial^2 s_3}{\partial t^2} = \frac{\partial \sigma_5}{\partial x_1} + \frac{\partial \sigma_4}{\partial x_2} + \frac{\partial \sigma_3}{\partial x_3}. \tag{8.50c}$$

Substituting σ_m from (8.49) and making use of the expression (3.12) for ε_m, the equations of motion (8.50) may be re-expressed as a single vector equation,

$$\rho_0 \frac{\partial^2 \boldsymbol{s}}{\partial t^2} = c_{44}^T \nabla^2 \boldsymbol{s} + (c_{12}^T + c_{44}^T) \text{ grad } \Delta - \mathscr{F} \text{ grad } T. \tag{8.51}$$

We now wish to incorporate the effects of heat flow into the equation of motion. Let dQ be the amount of heat entering owing to the thermal conductivity, χ, into an element of volume, per unit volume and in time dt,

$$\frac{\mathrm{d}Q}{\mathrm{d}t} = T \frac{\mathrm{d}S}{\mathrm{d}t} = \chi \nabla^2 T \tag{8.52}$$

where dS may be replaced by (8.49*a*) and hence one obtains

$$\chi \nabla^2 T = C_V \left(\frac{\partial T}{\partial t}\right) + T\mathscr{F}\left(\frac{\partial \Delta}{\partial t}\right). \tag{8.53}$$

For $\Delta = 0$, it is obvious that no temperature variation occurs due to the passage of the wave. Thus, shear waves propagate without incurring any attenuation due to thermal conductivity.

To calculate the attenuation of the plane dilatational wave we proceed as follows. Let us assume the solutions of equations (8.51) and (8.53) to be:

$$s_1 = s_0 \exp [\mathrm{i}(\omega t - k x_1)] \tag{8.54}$$

$$T - T_0 = T_1 \exp [\mathrm{i}(\omega t - k x_1)] \tag{8.55}$$

where T_0 is the mean temperature of the solid. Obviously for dilatational waves $s_2 = s_3 = 0$ and therefore $\Delta = -iks_1$. Substitution of (8.54) and (8.55) in equation (8.53) gives

$$T_0 - T = \frac{T\mathscr{F}\omega k s_1}{\chi k^2 + \mathrm{i}\omega C_V}. \tag{8.56}$$

Equations (8.54) to (8.56) are now substituted in equation (8.51) which, on making use of (8.43), (8.45) and (8.47), yields as the final relation between ω and k,

$$k^2 = \rho_0 \omega^2 \left(\frac{\omega C_V - \mathrm{i}k^2\chi}{\omega C_V M_S - \mathrm{i}k^2 \chi M_T}\right) \tag{8.57}$$

where

$$M_T = c_{12}^T + 2c_{44}^T = B_T + \tfrac{4}{3}G \tag{8.58}$$

$$M_S = M_T + (\gamma - 1)B_T = B_S + \tfrac{4}{3}G. \tag{8.59}$$

Equation (8.57) is a useful relation which will now be used to obtain the velocity and absorption of a plane wave. For this purpose we choose a complex representation† of the wavevector k, i.e.

$$k = k_1 - \mathrm{i}k_2 \qquad k_2 \geqslant 0 \tag{8.60}$$

where the real part, k_1, determines the phase velocity $C = \omega/k_1$ and the imaginary part, k_2, is a measure of absorption or attenuation. Substituting (8.60) into (8.54) one finds that the amplitude of the wave diminishes to e^{-1} of its value in a distance $x_1 = 1/k_2$. If the distance is measured in centimetres then k_2 is known as the amplitude attenuation per centimetre or simply the attenuation or absorption per centimetre. Usually, the distances are measured in units of wavelength and so the amplitude attenuation per wavelength is defined as

$$\alpha = k_2\lambda = 2\pi k_2/k_1. \tag{8.61}$$

Furthermore, equation (8.60) may also be written as

$$\begin{aligned} k^2 &= k_1^2 - k_2^2 - 2\mathrm{i}k_1k_2 \\ &\simeq k_1^2 - 2\mathrm{i}k_1k_2. \end{aligned} \tag{8.62}$$

The last equality follows because $k_2^2/k_1^2 \ll 1$, which will become clear shortly. Substituting (8.62) into equation (8.57) and on eliminating small-order terms, one obtains,

$$C^2 = C_0^2 \frac{f^2 + f_\chi^2}{\mathscr{D}f^2 + f_\chi^2} \qquad C_0^2 = \frac{M_S}{\rho_0} \tag{8.63}$$

$$\alpha = \pi\left(\frac{C}{C_0}\right)^2 (\mathscr{D} - 1)\frac{ff_\chi}{f^2 + f_\chi^2} \qquad \mathscr{D} = \frac{M_S}{M_T} \tag{8.64}$$

where

$$f = \frac{\omega}{2\pi} \qquad \text{and } f_\chi = \frac{\mathscr{D}C_VC^2}{2\pi\chi} \tag{8.65}$$

which are the required relations for the velocity, C, and for the attenuation, α, per wavelength for the plane dilatational waves. It may be noticed that the velocity of the dilatational wave is determined by the adiabatic elastic

† i in equation (8.60) must not be confused with the index i as used earlier. Here it stands for $\mathrm{i} = \sqrt{-1}$.

constants ($C^2 = C_0^2 = M_S/\rho_0$) at low frequency $f \ll f_\chi$, and by the isothermal elastic constants ($C_\infty^2 = M_T/\rho_0$) at high frequency $f \gg f_\chi$.

From equation (8.64) it is evident that the attenuation α is frequency dependent. The maximum value of α occurs at $f \simeq f_\chi$,

$$\alpha_{max} \simeq \pi \left(\frac{\mathscr{D} - 1}{\mathscr{D} + 1} \right). \tag{8.66}$$

At room temperature most solids have $\mathscr{D} \simeq 1.05$, hence $\alpha_{max} \sim 0.8$. At low frequency ($f \ll f_\chi$)

$$\alpha \simeq \pi(\mathscr{D} - 1)f/f_\chi \tag{8.67}$$

where f_χ can be obtained from (8.65). For a metal at room temperature a typical value of $f_\chi \sim 10^{10}$ to 10^{12} Hz. Since $\mathscr{D} - 1 \sim 10^{-2}$ the attenuation α therefore remains small.

Problems

8.1 Show that the elastic compliance constants of cubic materials satisfy the relations,

$$s_{11}^S - s_{11}^T = s_{12}^S - s_{12}^T = -\alpha^2 T/9C_P \tag{P8.1a}$$

$$s_{44}^S - s_{44}^T = 0 \tag{P8.1b}$$

where α is the coefficient of volume expansion.

8.2 For crystals obeying Hooke's law, the entropy to a good approximation can be regarded as a function of temperature only and to be independent of the strain variables. Show that: (*a*) The Helmholtz free energy for such a crystal with cubic symmetry has the form,

$$A(T) = A_0(T) + \tfrac{1}{2}c_{11}^T(\varepsilon_1^2 + \varepsilon_2^2 + \varepsilon_3^2) + \tfrac{1}{2}c_{44}^T(\varepsilon_4^2 + \varepsilon_5^2 + \varepsilon_6^2)$$
$$+ c_{12}^T(\varepsilon_1\varepsilon_2 + \varepsilon_2\varepsilon_3 + \varepsilon_3\varepsilon_1) \tag{P8.2}$$

where $A_0(T)$ is a function of the temperature only.

(*b*) How is $A_0(T)$ related to entropy?

(*c*) What are the values of the coefficients of thermal stress and strain for such crystals?

9

APPLICATION OF EQUILIBRIUM EQUATION TO FLUIDS

The physics of solids was the major subject of the discussion in the preceding chapters. It is the purpose of the remaining chapters to show that the formalism developed for the case of solids can be useful for fluids also. Fluids form the part of the deformable media which, unlike solids, flow easily with the application of stress. Fluids comprise both liquids and gases. Gases are highly compressible whereas liquids are virtually incompressible. Our common experience, however, shows that there is an overlap between the behaviour of liquids and gases. With a view to avoiding confusion we prefer to use the nomenclature of incompressible fluid (where no change in density occurs due to pressure) and compressible fluid (density is no longer considered constant) for our discussion. Homogeneous, incompressible and frictionless (inviscid) fluids shall be referred to as the ideal or perfect fluid. The added effect of viscosity makes it a real fluid.

Since fluids at rest cannot sustain shear stress ($\sigma_{ij}=0,\ i\neq j$), there are only normal components of stress or pressure. This greatly simplifies the form of the equilibrium equation. We shall see how the general form of the equilibrium equation derived earlier (see equation (4.8)),

$$\frac{\partial \sigma_{ji}}{\partial x_j}+\rho F_i=0$$

reduces to a simpler form for static fluids. In the present chapter its applications are intended both for incompressible and compressible fluids.

9.1 Equilibrium equation for fluids at rest

In a fluid at rest the force acting on any surface element within the fluid is always normal to the surface and is directed inwards (see figure 9.1). Such a force acting on a unit surface area is called fluid pressure and has the same value in all directions, $p=p_{x_1}=p_{x_2}=p_{x_3}$. Because of this isotropy, σ_{ij} must be diagonal with all diagonal elements equal,

$$\sigma_{ij}=-p\delta_{ij}=\begin{pmatrix}-p & 0 & 0\\ 0 & -p & 0\\ 0 & 0 & -p\end{pmatrix}. \tag{9.1}$$

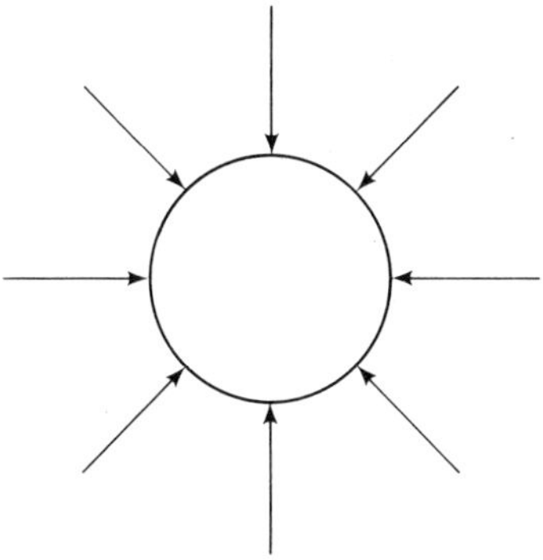

Figure 9.1 The forces acting on a surface element inside a static fluid.

Substituting this into the equilibrium equation (4.8), one has

$$-\frac{\partial p}{\partial x_i}+\rho F_i=0$$

or

$$\rho^{-1}(\text{grad } p)_i=F_i. \tag{9.2}$$

Since, in practice, the density ρ depends on p (p and ρ are single-valued functions), we therefore introduce a function $\mathscr{P}$ such that

$$\mathrm{d}\mathscr{P}=\mathrm{d}p/\rho \tag{9.3a}$$

or

$$\mathscr{P}=\int_A^B \frac{\mathrm{d}p}{\rho}. \tag{9.3b}$$

The limits of integration vary from a fixed reference point A to a variable field point $B(\boldsymbol{x})$. The integration in equation (9.3) can be evaluated once the relationship between p and ρ is known. Now, by virtue of the vector identity $\mathrm{d}\mathscr{P}\equiv\nabla\mathscr{P}\cdot\mathrm{d}\boldsymbol{r}$ and $\mathrm{d}p\equiv\nabla p\cdot\mathrm{d}\boldsymbol{r}$, we express equation (9.3a) as

$$\text{grad } \mathscr{P}=\rho^{-1} \text{ grad } p. \tag{9.4}$$

Hence, equation (9.2) becomes

$$\text{grad } \mathscr{P}=\boldsymbol{F}. \tag{9.5}$$

Equilibrium is thus only possible if the external force per unit mass is a conservative force, i.e. $\boldsymbol{F}$ must be derivable from the potential, say, U, i.e.

$$\boldsymbol{F}=-\text{grad } U. \tag{9.6}$$

Taking equations (9.5) and (9.6) together,

$$\text{grad}(\mathscr{P}+U)=0$$

or

$$\mathscr{P}+U=\text{constant}. \tag{9.7}$$

Equation (9.7) describes the equilibrium of fluids at rest.

9.2 Incompressible fluids

9.2.1 Pressure at different points in a liquid due to gravity

The external forces acting on a liquid are usually weak compared with the force required for significant compression and thus ρ may be regarded as a constant. This allows one to write the function $\mathscr{P}=p/\rho$ and hence equation (9.7) becomes

$$p\rho^{-1}+U=\text{constant}. \tag{9.8}$$

As ρ is constant, the surfaces of equal pressure are equipotential surfaces. Let us consider U due to gravity in a liquid at a depth z vertically beneath the free surface $z=0$,

$$U=-gz. \tag{9.9}$$

With this equation (9.8) becomes

$$p-\rho gz=\text{constant}. \tag{9.10}$$

At $z=0$, $p=p_0=$atmospheric pressure. Therefore,

$$p-p_0=\rho gz. \tag{9.11}$$

This is the well-known result for communicating vessels: (i) At equal depths there is an equal pressure in every part of the interconnected vessels regardless of their shapes and sizes. (ii) For a given pressure the height required for different liquids is different. That is why the atmospheric pressure corresponds to 76 cm of mercury but 76×13.6 cm of water (density of mercury= 13.6 g cm^{-3} and of water $\simeq 1.0$ g cm^{-3}).

9.2.2 Free surface of a liquid in a rotating drum

Consider a liquid in a drum which is rotating with angular velocity ω about its z-axis. The liquid as a whole rotates with the container and maintains a state of equilibrium. The resulting free surface (liquid–air interface) of the liquid is not a flat horizontal but rather is curves as shown in figure 9.2. The curved surface is an attribute of the combined effect of gravity and the centrifugal force. At a distance r ($0\leqslant r\leqslant a$, a the radius of the drum) from the axis, the centrifugal force per unit mass has the magnitude$=v^2/r=r\omega^2$ and is directed away from

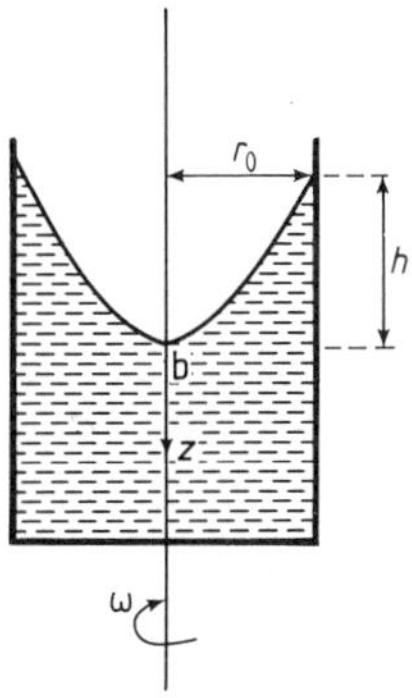

Figure 9.2 The free surface of a liquid in a rotating drum.

the axis. Therefore, the potential is

$$U = -gz - \tfrac{1}{2}r^2\omega^2. \tag{9.12}$$

The equilibrium equation (9.8) becomes

$$p + \rho(-gz - \tfrac{1}{2}r^2\omega^2) = \text{constant}. \tag{9.13}$$

At the point b (see figure 9.2), $r=0$, $p=p_0$ and let $z=z_0$. These determine the constant in equation (9.13) which allows us to write,

$$p - p_0 = \rho g(z - z_0 + r^2\omega^2/2g). \tag{9.14}$$

For the free surface, $p=p_0$, equation (9.14) reduces to

$$z_0 - z = r^2\omega^2/2g \tag{9.15}$$

which represents a paraboloid of rotation. The shape of the paraboloid or free surface obviously depends upon the angular velocity ω. It may be noticed that all the points at equal depths from the free surface are at the same pressure and thus they form congruent paraboloids.

Since $r_{max} = a$ (the radius of the drum) the rise of liquid from its vertex along the wall of the drum can be obtained from equation (9.15) as,

$$h = \frac{a^2\omega^2}{2g} = \frac{v^2}{2g}. \tag{9.16}$$

This is also known as the velocity head. $v = a\omega$ is the circumferential velocity.

9.3 Compressible fluids

9.3.1 Equilibrium equation for isothermal case

Unlike an incompressible fluid, the density of a compressible fluid is not independent of pressure. Before we can rewrite the equilibrium equation (9.7) for compressible fluids, one requires a suitable relation between p and ρ in order to evaluate $\mathscr{P}$. One of the simplest relations is the equation of state for an ideal gas,

$$pV = NRT \tag{9.17}$$

or

$$p = \rho RT/M \qquad \text{with } \rho = NM/V \tag{9.18}$$

where R is the universal gas constant, N is Avogadro's number and M is the molecular weight. At constant temperature (isothermal state) equation (9.18) may be written as

$$p = c\rho \tag{9.19}$$

where c is a constant. Thus the function $\mathscr{P}$ becomes (see equation (9.3))

$$\mathscr{P} = \frac{RT}{M} \int \frac{\mathrm{d}p}{p}$$

or

$$\mathscr{P} = \frac{RT}{M} \ln p. \tag{9.20}$$

Consider now the problem of the rotating drum as discussed in §9.2.2 and assume that the temperature T of the containing compressible fluid is the same everywhere. Equivalent to equation (9.12), the potential U can be written as

$$U = gz - \tfrac{1}{2} r^2 \omega^2 \tag{9.21}$$

where z is counted positive upwards. Substituting the values of $\mathscr{P}$ and U in the equilibrium equation (9.7), one has

$$\ln p + (M/RT)(gz - \tfrac{1}{2} r^2 \omega^2) = \text{constant}. \tag{9.22}$$

At $z = 0$ and $r = 0$, the pressure $p = p_0$ and hence in equation (9.22) the constant is $\ln p_0$. It can now be written as

$$p = p_0 \exp\,[(M/RT)(-gz + \tfrac{1}{2} r^2 \omega^2)]. \tag{9.23}$$

Thus the surfaces of constant pressure are paraboloids. $p = p_0$ is the pressure of a particular paraboloid namely,

$$-gz + \tfrac{1}{2} r^2 \omega^2 = 0. \tag{9.24}$$

The paraboloids with $p < p_0$ and $p > p_0$ lie correspondingly above and below the reference paraboloid (9.24) as shown in figure 9.3. Now we give some physical applications of equation (9.23).

(i) Principle of the centrifuge

Neglecting the effect of gravity ($g = 0$), equation (9.23) reduces to

$$p = p_0 \exp\,[(M/2RT) r^2 \omega^2]. \tag{9.25}$$

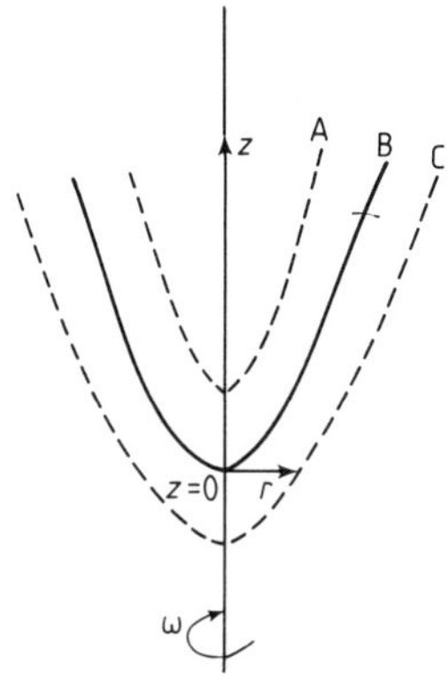

Figure 9.3 The rotation of a compressible fluid in a circular drum. The curves are: A, $p<p_0$;B, $p=p_0$; C, $p>p_0$.

At $r=0$, one has $p=p_0$. On account of the exponential factor in equation (9.25), the pressure increases quickly with increasing r. For a given r and ω, the pressure also depends upon the molecular weight M. If we take a mixture of the compressible fluids, say two gases of molecular weights M_1 and M_2, then following equation (9.25) the partial pressures p_i can be written as,

$$p_i=p_{0i}\exp\,[(M_i/2RT)r^2\omega^2] \qquad i=1, 2 \tag{9.26}$$

where p_{01} and p_{02} are the partial pressures of two gases at $r=0$. The total pressure is obtained by linear superpositions, $p=p_1+p_2$. It is evident from (9.26) that the pressure of the heavier-mass component increases faster than that of the lower-mass component with increasing r. This results in a concentration of the heavier-mass component at the edges and the lower-mass component near the axis of the drum. This is known as the principle of the centrifuge. Since its discovery the commercialised applications have been developed and it is utilised extensively, for instance, in the separation of cream from milk, the partial separation of gases in the terrestrial atmosphere, the separation of uranium isotopes and in many other applications. A brief history and the various uses of the principle of the centrifuge is given in the recent review by Stanley (1984). Even the science of biology has benefited to a great extent from the principle of the centrifuge where it is used to separate a mixture consisting of heavy protein molecules (molecular weight $\sim$30 000).

(ii) Barometric formula for an isothermal atmosphere

For $\omega=0$, equation (9.23) becomes

$$p=p_0\exp\,(-Mgz/RT). \tag{9.27}$$

This can be used to determine the height, z, of a point in the atmosphere above sea level directly from a barometer reading. For instance, consider the atmosphere:

$$M=29\times 10^{-3}\ \mathrm{kg} \qquad R=8.3\ \mathrm{J\,K^{-1}}$$

$$T=273\ \mathrm{K} \qquad g=9.8\ \mathrm{m\,s^{-2}} \qquad z=1000\ \mathrm{m}$$

$$\frac{p}{p_0}=\exp\left(\frac{-29\times 10^{-3}\times 9.8\times 10^3}{8.3\times 273}\right)\simeq 0.88.$$

The calculation shows that the barometer reading $p/p_0\simeq 0.88$ at a point in an isothermal atmosphere corresponds to an altitude of 1000 m.

For the state of constant temperature equation (9.27) can equivalently be written in terms of density,

$$\rho=\rho_0 \exp(-Mgz/RT) \tag{9.28}$$

where ρ_0 is the density at the reference level. Equation (9.28) permits us to calculate the density distribution. If we write $Mgz/RT=mgz/kT=U/kT$ (U is the potential energy of the mass point m in the field of gravity g at an altitude z) then (9.28) leads to

$$\rho=\rho_0 \exp(-U/kT). \tag{9.29}$$

The exponential function of (9.29) is well known to us as the Boltzmann factor, a fundamental quantity in the statistical investigations.

9.3.2 Temperature distribution in a polytropic atmosphere

The assumption of constant T as made in the preceding examples is, in general, not valid over the entire range of atmospheric heights. If this is the case then the pressure–density relation $p=c\rho$ (see equation (9.19)) no longer holds true. For a polytropic atmosphere one therefore writes,

$$p=c\rho^n \tag{9.30}$$

where n is called the polytropic exponent and c is a constant. $n=1$ corresponds to the isothermal state. For the adiabatic state n depends upon the type of gas, for instance, $n=\frac{5}{3}$ for monatomic gases and $n=\frac{7}{5}$ for diatomic gases. However, $n=\frac{6}{5}$ is chosen very often for most aeronautical purposes.

The constant c in equation (9.30) can be determined in terms of the reference pressure p_0 and density ρ_0, i.e.

$$c=p_0/\rho_0^n. \tag{9.31}$$

The function $\mathscr{P}$, therefore, becomes

$$\mathscr{P}=\int\frac{\mathrm{d}p}{\rho}=\frac{p_0^{1/n}p^{(1-1/n)}}{\rho_0(1-n^{-1})}. \tag{9.32}$$

Now substitute $\mathscr{P}$ and $U=gz$ in the equilibrium equation (9.7),

$$\frac{p_0^{1/n}p^{(1-1/n)}}{\rho_0(1-n^{-1})}+gz=\text{constant}. \tag{9.33}$$

At $z=0$, $p=p_0$ (sea level pressure):

$$\text{constant}=p_0/\rho_0(1-n^{-1}). \tag{9.34}$$

Therefore equation (9.33) reduces to

$$p=p_0\left(1-gz\frac{\rho_0}{p_0}\frac{(n-1)}{n}\right)^{n/(n-1)}. \tag{9.35}$$

This is the general expression for the pressure in a polytropic atmosphere. Denoting the effective height of the atmosphere by

$$h=\frac{np_0}{(n-1)\rho_0 g}=\frac{n}{n-1}\frac{RT_0}{Mg} \tag{9.36}$$

one has

$$p=p_0(1-zh^{-1})^{n/(n-1)}. \tag{9.37}$$

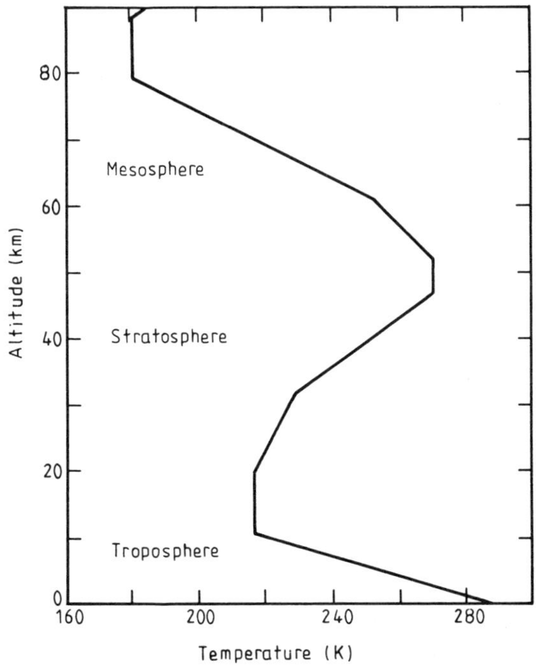

Figure 9.4 The temperature distribution in atmospheric layers. (Adopted from R C Haymes, 1971, 'Introduction to Space Science' (New York: John Wiley) p 56.)

Since $p/p_0=(\rho/\rho_0)^n$, equation (9.37) can also be expressed as

$$\rho=\rho_0(1-zh^{-1})^{1/(n-1)}. \tag{9.38}$$

Making use of equations (9.37) and (9.38) with the equation of state (9.18), the distribution of temperature in a polytropic atmosphere is:

$$T=T_0(1-z/h). \tag{9.39}$$

Thus the temperature drops linearly with altitude in a polytropic atmosphere. It is, however, very important to realise that the relation (9.39) is given here only to bring out the essential features and should not be considered as a rigorous theoretical expression for this problem. In reality the temperature distribution in various atmospheric layers is quite complex (for instance, see figure 9.4) and by no means could be explained from (9.39) uniformly. In the lowest part of the atmosphere, which is called the troposphere, the temperature drops with altitude up to about 12 km. The decreasing temperature terminates at the tropopause where it remains almost constant for about 5 km and then increases with altitude in the stratosphere. The temperature starts decreasing again in the region of the mesophere.

Problems

9.1 The density of water (in $\mathrm{g\,cm^{-3}}$) at moderate pressures varies as

$$\rho=1+bp \qquad b=5\times 10^{-11}\ \mathrm{g\,cm^{-1}\,dyn^{-1}}. \tag{P9.1}$$

(i) If p_0 is the atmospheric pressure at the surface of a lake, show that the pressure of water at depth h in the lake is approximately given by

$$p-p_0=b^{-1}(\mathrm{e}^{bgh}-1). \tag{P9.2}$$

In deriving (P9.2) a term bp_0 has been neglected compared with unity. Is this neglect justified?

(ii) What is the value of $p-p_0$ at a depth of 1000 m?

9.2 If the surface of the earth is considered to be spherical rather than planar, then show that the expression for the pressure variation in a polytropic atmosphere becomes

$$p=p_0\left[1-G_0\mathscr{M}\left(\frac{1}{R}-\frac{1}{r}\right)\frac{\rho_0}{p_0}\frac{n-1}{n}\right]^{n/(n-1)}. \tag{P9.3}$$

Here G is the constant of Newton's law of gravitation, $\mathscr{M}$ is the mass of the earth, R is the radius of the earth and r is the distance from the centre of the earth. Does the pressure vanish as $r\to\infty$ in the isothermal atmosphere?

10

PERFECT FLUIDS: EULER'S EQUATIONS

In the previous Chapter we briefly discussed the static (time-independent) properties of fluids. We now investigate the flow problems in perfect fluids (inviscid or frictionless). As in Chapter 6, the spatial description for particles shall be considered. We recall the velocity flow function $\boldsymbol{v}(\boldsymbol{x}, t)$ which is the velocity of an element of fluid at time t and at a point $\boldsymbol{x}$ fixed in space ($\boldsymbol{x}$ and t are independent variables). If the velocity at a given point $\boldsymbol{x}$ is independent of time then the flow is called steady flow, otherwise it is unsteady flow. In addition to $\boldsymbol{v}(\boldsymbol{x}, t)$ flow problems also require a knowledge of the density of the fluid at a point $\boldsymbol{x}$ and at time t which is denoted here as $\rho(\boldsymbol{x}, t)$. The density is also a function of both $\boldsymbol{x}$ and t unless the fluid is incompressible and homogeneous in which case $\rho(\boldsymbol{x}, t)$ is constant in time as well as in space.

The present Chapter introduces some basic equations such as the equation of continuity, Euler's equation of motion and Bernoulli's equation together with some simple applications.

10.1 Equation of continuity

The conservation of matter requires that the equation of continuity for fluids in motion is satisfied. The variations of the variables ρ $(\equiv \rho(\boldsymbol{x}, t))$ and $\boldsymbol{v}$ $(\equiv \boldsymbol{v}(\boldsymbol{x}, t))$ are not independent of each other. To reach the required relationship consider a volume V_0 fixed in space (also called the control volume in the terminology of fluid dynamics) which is occupied with the fluid and is bounded by a closed surface A as shown in figure 10.1. The mass of fluid flowing out per unit time through an elemental area $\mathrm{d}A$ is $\rho \boldsymbol{v} \cdot \boldsymbol{n}\, \mathrm{d}A$, where $\boldsymbol{n}$ is the unit vector normal to $\mathrm{d}A$ and is directed outwards. The total mass flowing out per unit time through the closed surface A is therefore,

$$\int_A (\rho \boldsymbol{v}) \cdot \boldsymbol{n}\, \mathrm{d}A. \tag{10.1}$$

This creates a depletion of fluid mass inside the volume V_0. The depletion per unit time may be written as

$$-\int_{V_0} \frac{\partial \rho}{\partial t}\, \mathrm{d}V \tag{10.2}$$

where the integration is taken over the volume V_0. From the conservation of

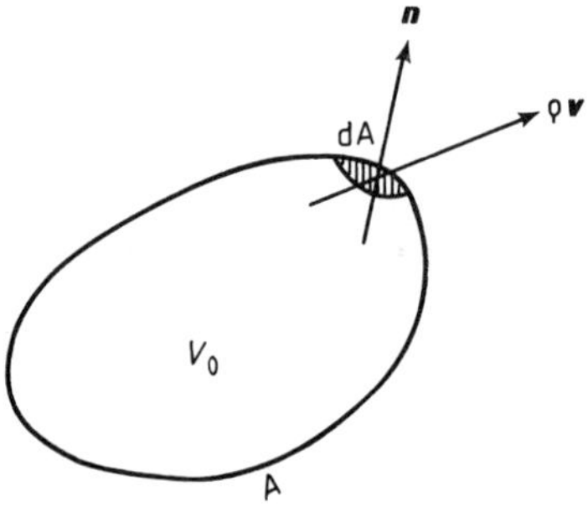

Figure 10.1.

mass one must have that

$$\int_A (\rho \boldsymbol{v})\cdot\boldsymbol{n}\,\mathrm{d}A = -\int_{V_0} \frac{\partial \rho}{\partial t}\,\mathrm{d}V. \tag{10.3}$$

Making use of the divergence theorem,

$$\int_A (\rho \boldsymbol{v})\cdot\boldsymbol{n}\,\mathrm{d}A = \int_{V_0} \operatorname{div}(\rho \boldsymbol{v})\,\mathrm{d}V \tag{10.4}$$

equation (10.3) becomes

$$\int_{V_0} \left(\frac{\partial \rho}{\partial t} + \operatorname{div}(\rho \boldsymbol{v})\right)\mathrm{d}V = 0. \tag{10.5}$$

This is true for every volume, therefore,

$$\frac{\partial \rho}{\partial t} + \operatorname{div}(\rho \boldsymbol{v}) = 0. \tag{10.6a}$$

or, equivalently

$$\frac{\partial \rho}{\partial t} + \frac{\partial}{\partial x_i}(\rho v_i) = 0 \tag{10.6b}$$

which is known as the continuity equation. Rewriting the second term one has

$$\frac{\partial \rho}{\partial t} + v_i \frac{\partial \rho}{\partial x_i} + \rho \frac{\partial v_i}{\partial x_i} = 0. \tag{10.7}$$

In terms of the material derivative (see equation (6.5)) the above equation can also be expressed as,

$$\frac{\mathrm{D}\rho}{\mathrm{D}t} + \rho \frac{\partial v_i}{\partial x_i} = 0 \tag{10.8a}$$

or,

$$\frac{\mathrm{D}\rho}{\mathrm{D}t} + \rho \operatorname{div} \boldsymbol{v} = 0. \tag{10.8b}$$

For an incompressible homogeneous fluid (ρ is constant in space and time), the continuity equation (10.7) reduces to

$$\operatorname{div} \boldsymbol{v}=0. \tag{10.9}$$

Equation (10.9) is often called the condition of incompressibility. It is important to realise here that (10.9) also remains true for an incompressible but heterogeneous fluid (ρ varies in space) by virtue of equation (10.8). For a fluid such as oil in water, ρ does not change with the fluid motion, that is, the material derivative $\mathrm{D}\rho/\mathrm{D}t=0$ and hence (10.9) follows.

10.2 Equations of motion (Euler's equations)

10.2.1 Euler's equation for a perfect fluid

The ground work for setting up the equations of motion has already appeared in Chapter 6. We recall that the most general form of the equations of motion (see equation (6.8)) is

$$\rho \frac{\mathrm{D}v_i}{\mathrm{D}t}=\frac{\partial \sigma_{ij}}{\partial x_j}+\rho F_i \qquad i=1, 2, 3$$

where v_i are the components of velocity, σ_{ij} are the components of the stress tensor and F_i are the components of the body force acting on unit mass of the matter. It is the specification of σ_{ij} which makes the above equation suitable either for a solid or for a fluid.

For a non-viscous (frictionless) flow the stress components σ_{ij} can be written as

$$\sigma_{ij}=-p\delta_{ij}$$

where p should not be confused with hydrostatic pressure as in equation (9.1) but should rather be treated as a fourth unknown variable along with v_i. Hence the equations of motion for a perfect fluid are

$$\rho \frac{\mathrm{D}v_i}{\mathrm{D}t}=-\frac{\partial p}{\partial x_i}+\rho F_i. \tag{10.10}$$

This equation was first put forward by Leonhard Euler as early as 1755 and is now well known as Euler's equation of motion for perfect fluids. We shall see that (10.10) forms the basis of the study of fluid dynamics. Before we attempt its solution it is worthwhile to examine its structure in greater detail.

10.2.2 Note on structure and alternative form of Euler's equation

Expanding the material derivative $\mathrm{D}/\mathrm{D}t$ as in equation (6.5), Euler's equation becomes

$$\rho\left(\frac{\partial v_i}{\partial t}+\sum_j v_j\frac{\partial v_i}{\partial x_j}\right)=-\frac{\partial p}{\partial x_i}+\rho F_i. \tag{10.11}$$

In order to bring (10.11) to a useful form let us express

$$\sum_j v_j\frac{\partial v_1}{\partial x_j}=v_1\frac{\partial v_1}{\partial x_1}+v_2\frac{\partial v_1}{\partial x_2}+v_3\frac{\partial v_1}{\partial x_3}$$

$$=v_1\frac{\partial v_1}{\partial x_1}+v_2\frac{\partial v_2}{\partial x_1}+v_3\frac{\partial v_3}{\partial x_1}+v_2\left(\frac{\partial v_1}{\partial x_2}-\frac{\partial v_2}{\partial x_1}\right)+v_3\left(\frac{\partial v_1}{\partial x_3}-\frac{\partial v_3}{\partial x_1}\right)$$

$$=\frac{1}{2}\frac{\partial}{\partial x_1}\boldsymbol{v}^2-(\boldsymbol{v}\times\text{curl }\boldsymbol{v})_1. \tag{10.12}$$

Similarly for $v_j\,\partial v_2/\partial x_j$ and $v_j\,\partial v_3/\partial x_j$ and therefore in vector notation one has

$$\sum_j v_j\frac{\partial v_i}{\partial x_j}=\frac{1}{2}\frac{\partial\boldsymbol{v}^2}{\partial x_i}-(\boldsymbol{v}\times\text{curl }\boldsymbol{v})_i. \tag{10.13}$$

Equation (10.11) now readily reduces to

$$\rho\frac{\partial v_i}{\partial t}+\frac{1}{2}\rho\frac{\partial\boldsymbol{v}^2}{\partial x_i}-\rho(\boldsymbol{v}\times\text{curl }\boldsymbol{v})_i=-\frac{\partial p}{\partial x_i}+\rho F_i \tag{10.14}$$

or,

$$\rho\left(\frac{\partial\boldsymbol{v}}{\partial t}-\boldsymbol{v}\times\text{curl }\boldsymbol{v}\right)+\tfrac{1}{2}\rho\text{ grad }\boldsymbol{v}^2+\text{grad }p=\rho\boldsymbol{F} \tag{10.15}$$

which is a very convenient representation of Euler's equation of motion. It can easily be specialised in any curvilinear coordinate system just by substituting the respective values of curl and grad.

Equation (10.15) is non-linear in $\boldsymbol{v}$ and hence is difficult to solve. Its successful solution seems possible only in specialised situations as discussed below.

10.3 Bernoulli's equation

10.3.1 Integration of Euler's equation

Euler's equation of motion (10.15) may readily be integrated if we assume:

(i) The flow is irrotational, that is, the vorticity $\boldsymbol{\omega}$ (see equation (3.22)) is zero,

$$\boldsymbol{\omega}=\tfrac{1}{2}\text{ curl }\boldsymbol{v}=0. \tag{10.16}$$

Therefore, we can define a potential ϕ, known as the velocity potential, such that

$$v = -\text{grad}\ \phi. \tag{10.17}$$

(ii) The body force $\boldsymbol{F}$ is conservative,

$$\boldsymbol{F} = -\text{grad}\ U. \tag{10.18}$$

Under the above two conditions, equation (10.15) becomes

$$-\frac{\partial}{\partial t}\text{grad}\ \phi + \tfrac{1}{2}\text{grad}\ v^2 + \frac{1}{\rho}\text{grad}\ p + \text{grad}\ U = 0. \tag{10.19}$$

Introducing, as before (see equation (9.3)), a function $\mathscr{P}$

$$\mathrm{d}\mathscr{P} = \mathrm{d}p/\rho \qquad \text{grad}\ \mathscr{P} = \text{grad}\ p/\rho$$

equation (10.19) reduces to

$$\text{grad}(-\partial\phi/\partial t + \tfrac{1}{2}v^2 + \mathscr{P} + U) = 0. \tag{10.20}$$

The integration of (10.20) yields the generalised form of Bernoulli's equation:

$$-\frac{\partial\phi}{\partial t} + \tfrac{1}{2}v^2 + \int\frac{\mathrm{d}p}{\rho} + U = \text{constant}. \tag{10.21a}$$

If the fluid is incompressible then

$$\int\frac{\mathrm{d}p}{\rho} = \frac{p}{\rho}$$

the above equation becomes

$$-\frac{\partial\phi}{\partial t} + \tfrac{1}{2}v^2 + \frac{p}{\rho} + U = \text{constant}. \tag{10.21b}$$

For steady flow v or ϕ is invariant in time ($\partial\phi/\partial t = 0$) which allows us to write

$$\tfrac{1}{2}v^2 + p\rho^{-1} + U = \text{constant}. \tag{10.22}$$

This is generally known as Bernoulli's equation or the pressure equation for steady flow.

10.3.2 Discussion of Bernoulli's equation

The first term $\frac{1}{2}v^2$ in (10.22) represents the kinetic energy per unit mass, p/ρ is the potential energy associated with internal forces and the last term U represents the potential energy due to external forces. Bernoulli's equation therefore has the simple physical interpretation that the sum total of the kinetic and potential energies in an irrotational steady flow of an incompressible fluid is constant. The potential energy, U, usually arises due to gravitational forces, i.e. $U=gz$, hence one has

$$\tfrac{1}{2}v^2+p\rho^{-1}+gz=\text{constant}. \qquad (10.23)$$

Some applications of equation (10.23) follow:

(i) Flow in a horizontal pipe of variable cross section.
Consider that an incompressible liquid flows in a horizontal pipe of variable cross section with A_1 and A_2 $(A_1<A_2)$ as shown in figure 10.2. Since the pipe is horizontal (gz is constant), equation (10.23) becomes

$$\tfrac{1}{2}v^2+p\rho^{-1}=\text{constant}. \qquad (10.24)$$

Furthermore, for an incompressible fluid the volume flux through each cross section of the pipe is the same, i.e.

$$A_1v_1=A_2v_2 \qquad (10.25)$$

where v_1 and v_2 are the velocities of flow at cross sectional areas A_1 and A_2. Since $A_1<A_2, v_1>v_2$, the velocity in the narrow section of pipe is greater than that in the wide section. It follows immediately from equation (10.24) that the pressure of liquid is less in the narrow section than the wide section.

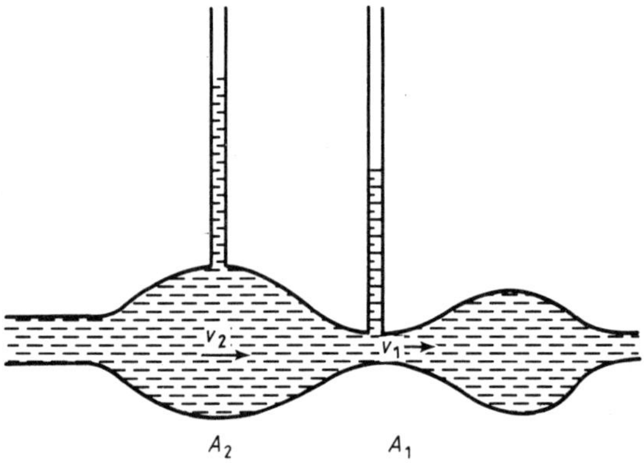

Figure 10.2 Flow of an incompressible fluid through a pipe of variable cross section.

(ii) Flow through an orifice at the bottom of a tank.
Consider a tank AB (figure 10.3) with a small orifice at the bottom which is filled with liquid. The pressure at the free surface ($z=0$) is p_A the atmospheric pressure and at the bottom, $z=-h$, let it be p_B. Applying the Bernoulli equation (10.23) at A and B one has

$$\tfrac{1}{2}v_A^2+p_A\rho^{-1}=\tfrac{1}{2}v_B^2+p_B\rho^{-1}-gh. \qquad (10.26)$$

v_A and v_B are the velocities at the free and bottom surfaces respectively. If the

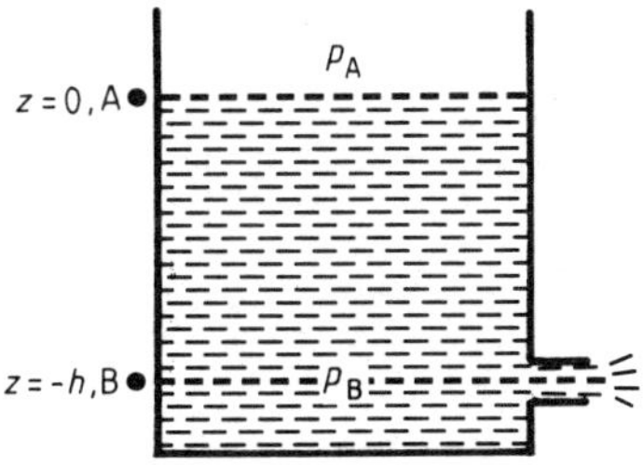

Figure 10.3 Flow through a small opening near the bottom of a tank.

orifice is closed, there is no motion, i.e. $v_A = v_B = 0$, and (10.26) yields

$$p_B - p_A = \rho g h \tag{10.27}$$

a standard result for hydrostatic pressure.

If, on the other hand, the orifice is opened, the free surface at A and the orifice at B are exposed to atmospheric pressure, i.e. $p_A = p_B$. The velocity at the free surface is zero, therefore (10.26) becomes

$$v_B^2 = 2\rho g h. \tag{10.28}$$

The relationship expressed by equation (10.28) is called Torricelli's theorem. We shall meet some more applications of Bernoulli's equation in due course.

10.4 Helmholtz–Kelvin theorem on vorticity

Before we look at other examples of irrotational flow, it is desirable to enunciate a theorem related to the rotational motion (usually defined in terms of the vorticity vector or vortices $\boldsymbol{\omega} = \frac{1}{2} \operatorname{curl} \boldsymbol{v}$, see equation (3.22)) of a fluid. It was Helmholtz who stated this theorem as early as 1858 and it was subsequently simplified by Lord Kelvin. The theorem states: (i) vortices can neither be created nor destroyed and (ii) the circulation (also called the vortex strength†) is constant in time for an inviscid and incompressible fluid. More explicitly, if a fluid motion is irrotational at a given instant of time then it remains so for all time and hence can be expressed in terms of the velocity potential.

To prove this, consider a necklace of fluid particles along a closed path ABCDA (called $\mathscr{C}$) lying well inside the fluid (figure 10.4). As the fluid moves the necklace of particles also moves. The circulation, Γ, associated with the closed path $\mathscr{C}$ is defined by the line integral,

$$\Gamma = \oint_{\mathscr{C}} \boldsymbol{v} \cdot \mathrm{d}\boldsymbol{s}. \tag{10.29}$$

† We shall discuss it in §12.5.

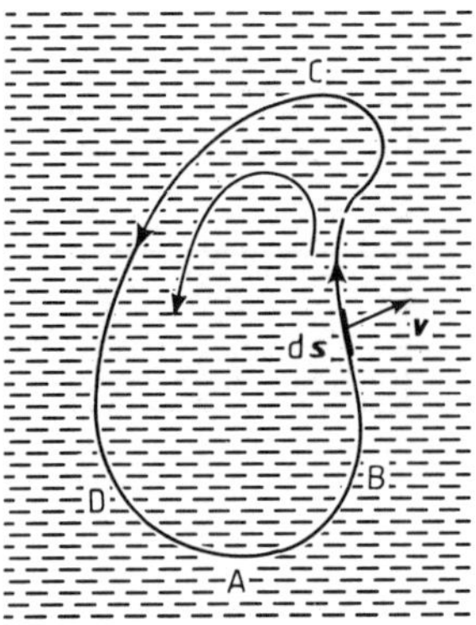

Figure 10.4 A necklace ABCDA of fluid particles inside a fluid.

Γ is positive if the circulation of the particles along $\mathscr{C}$ is counterclockwise and negative otherwise. Equation (10.29) can be expanded:

$$\Gamma = \oint_{\mathscr{C}} \left(v_1 \frac{dx_1}{ds} + v_2 \frac{dx_2}{ds} + v_3 \frac{dx_3}{ds} \right) ds$$

$$= \oint_{\mathscr{C}} (v_1 \, dx_1 + v_2 \, dx_2 + v_3 \, dx_3). \tag{10.30}$$

The rate of change of Γ with time is obtained using the material derivative,

$$\frac{D\Gamma}{Dt} = \frac{D}{Dt} \oint \boldsymbol{v} \cdot d\boldsymbol{s}$$

$$= \oint \left(\frac{D\boldsymbol{v}}{Dt} \right) \cdot d\boldsymbol{s} + \int \boldsymbol{v} \cdot \frac{D}{Dt} \, d\boldsymbol{s}. \tag{10.31}$$

If we express

$$\frac{D}{Dt} \, d\boldsymbol{s} = d \left(\frac{D\boldsymbol{s}}{Dt} \right) = d\boldsymbol{v}$$

then equation (10.31) becomes

$$\frac{D\Gamma}{Dt} = \oint \left(\frac{D\boldsymbol{v}}{Dt} \right) \cdot d\boldsymbol{s} + \oint \boldsymbol{v} \cdot d\boldsymbol{v}. \tag{10.32}$$

Obviously the last term in equation (10.32) is zero, therefore one has

$$\frac{D\Gamma}{Dt} = \oint \left(\frac{D\boldsymbol{v}}{Dt} \right) \cdot d\boldsymbol{s} \tag{10.33}$$

where $D\boldsymbol{v}/Dt$ may be obtained from Euler's equation (10.10),

$$\rho \frac{D\boldsymbol{v}}{Dt} = -\text{grad } p + \rho \boldsymbol{F}.$$

For conservative forces we may introduce, as we did earlier, the functions; grad $\mathscr{P}$ = grad p/ρ (equation (9.4)) and $\boldsymbol{F} = -\text{grad } U$ (equation (9.6)). These substitutions make it possible to write

$$\frac{\mathrm{D}\Gamma}{\mathrm{D}t} = \oint(-\text{grad } \mathscr{P} - \text{grad } U)\cdot \mathrm{d}\boldsymbol{s}. \tag{10.34}$$

Since U and $\mathscr{P}$ are single-valued functions of the space coordinates only,

$$\mathrm{D}\Gamma/\mathrm{D}t = 0 \qquad \Gamma = \text{constant}. \tag{10.35}$$

This relation is the conservation law of circulation. Γ is independent of t, i.e. whatever vortices are present at any time it remains the same for all times. It may, however, be noticed that the above result is true only for an inviscid fluid under conservative forces. For a viscous liquid there is damping and the vortices die out.

Using Stoke's theorem, equation (10.29) can also be expressed as

$$\Gamma = \oint_{\mathscr{C}} \boldsymbol{v}\cdot \mathrm{d}\boldsymbol{s} = \int_A \text{curl } \boldsymbol{v}\cdot \mathrm{d}\boldsymbol{A} \tag{10.36}$$

which shows that if curl $\boldsymbol{v} = 0$ (i.e. the vorticity $\boldsymbol{\omega}$ is zero) everywhere in the region of fluid then for every closed curve lying in the fluid $\Gamma = 0$. Hence by virtue of (10.35) the flow is permanently irrotational.

Problems

10.1 Consider the motion of a liquid in a cylinder where the axis is taken along the x_3-axis. If the motion starts from rest under the action of the forces

$$F_1 = \alpha x_1 + \beta x_2 \qquad F_2 = \alpha' x_1 + \beta' x_2 \qquad \text{and } F_3 = 0 \tag{P10.1}$$

then show that

$$p\rho^{-1} = \tfrac{1}{2}\omega^2(x_1^2 + x_2^2) + \tfrac{1}{2}[\alpha x_1^2 + (\beta + \beta')x_1 x_2 + \beta' x_2^2] + \text{constant}. \tag{P10.2}$$

where ω is the angular velocity and is a function of t only.

10.2 A gas, in which p and ρ are related by an adiabatic relation $p = c\rho^n$, flows in a steady state in a horizontal pipe. Using Bernoulli's equation show that an upper limit to the value of the flow velocity is given by

$$v_{\text{max}} = c_0\left(\frac{2}{n-1}\right)^{1/2} \tag{P10.3}$$

where $c_0 = (np_0/\rho_0)^{1/2}$ is the velocity of sound; p_0 and ρ_0 refer to the standard state.

11

IRROTATIONAL (POTENTIAL FLOW) IN THREE DIMENSIONS

Having established the basic equations of hydrodynamics, we now specialise these to some specific irrotational flow problems. Flow which is irrotational everywhere (except possibly at a few singular points or lines) is also known as potential flow. At the beginning of this Chapter the problem of the potential flow of a perfect incompressible fluid is formulated leading to Laplace's equation. The concepts of sources, sinks and the method of images are briefly introduced. The remaining part of this Chapter is devoted to the solution of Laplace's equation for the irrotational motion produced in a liquid by the uniform and the accelerated motions of a sphere. We shall see how this result can be used to obtain useful information, for instance, the force exerted on a sphere. We restrict our discussion to three dimensions in this Chapter and to two dimensions in the next Chapter.

11.1 Laplace's equation and boundary conditions

11.1.1 Laplace's equation

In the case of irrotational motion, curl $\boldsymbol{v}=0$ which implies that $\boldsymbol{v}$ can be expressed in terms of the velocity potential ϕ (see equation (10.17)):

$$\boldsymbol{v}=-\text{grad}\,\phi.$$

The condition of incompressibility (equation (10.9)), on the other hand, demands that

$$\text{div}\,\boldsymbol{v}=0.$$

Combining the above equations, one has

$$\text{div grad}\,\phi=0$$

or

$$\nabla^2\phi=0 \qquad (11.1a)$$

where ∇^2 is called the Laplacian operator and equation (11.1*a*) is known as

Laplace's equation. For Cartesian axes (x_1, x_2, x_3),

$$\nabla^2 \equiv \frac{\partial^2}{\partial x_1^2} + \frac{\partial^2}{\partial x_2^2} + \frac{\partial^2}{\partial x_3^2}. \tag{11.1b}$$

For cylindrical polar coordinates (r, φ, z),

$$\nabla^2 = \frac{1}{r}\left[\frac{\partial}{\partial r}\left(r\frac{\partial}{\partial r}\right) + \frac{1}{r}\frac{\partial^2}{\partial \varphi^2} + r\frac{\partial^2}{\partial z^2}\right] \tag{11.1c}$$

and for spherical polar coordinates (r, θ, φ)

$$\nabla^2 = \frac{1}{r^2 \sin\theta}\left[\sin\theta \frac{\partial}{\partial r}\left(r^2 \frac{\partial}{\partial r}\right) + \frac{\partial}{\partial \theta}\left(\sin\theta \frac{\partial}{\partial \theta}\right) + \frac{1}{\sin\theta}\frac{\partial^2}{\partial \varphi^2}\right]. \tag{11.1d}$$

Equation (11.1*a*) is true everywhere for irrotational flow except at singular points or lines. Remembering Laplace's equation of electrostatics ($\nabla^2\phi_{ES} = 0$, except where charges may be present), one recognises that the flow problems in fluids are very similar to problems in electrostatics.

As there are an infinite number of solutions of Laplace's equation, the principal concern of this Chapter is to make a proper choice of function ϕ which has a physical meaning subject to the appropriate boundary conditions. On account of the similarity of this problem with electrostatics, we can make use of the solutions already known from, for instance, the solution for one or more point charges. Corresponding to a point charge one has a source or a sink in hydrodynamics. We consider simple examples such as those in the following section.

11.1.2 Boundary conditions for perfect fluid

In order to choose a physically acceptable solution of Laplace's equation, it is important to know the flow and boundary conditions at the surfaces of solid objects which the fluid passes. For a perfect fluid these boundary conditions are: (i) The normal component of the fluid velocity must be zero at the surface of stationary solid objects in the fluid. This derives from the fact that the fluid cannot penetrate into the solid boundary. (ii) For moving solid objects the component of the fluid velocity along the surface normal is equal to the component of the solid velocity along the surface normal.

11.2 Some simple examples of irrotational flow

11.2.1 Sources and sinks

We have emphasised that the well-known solutions of Laplace's equation in electrostatics are of great help in describing the irrotational flow of fluids. In analogy with the electrostatic potential ($\phi_{ES}=(e/4\pi\varepsilon_0)r^{-1}$ we express the velocity potential ϕ as

$$\phi=A/r \qquad A=m/4\pi \tag{11.2}$$

where A is a constant and r denotes the distance from a point $\circ$. From this the radial velocity of the flow, v_r, may be obtained.

$$v_r=-\frac{\partial\phi}{\partial r}=\frac{m}{4\pi r^2}. \tag{11.3}$$

The total flow per unit time across a circle of radius r about $\circ$, is therefore,

$$4\pi r^2 v_r=m. \tag{11.4}$$

Thus m is defined as the amount of fluid which appears per unit time at the point $\circ$. For this reason the point $\circ$ is known as a source and m is the measure of strength of the source. Any flow originating from a source is always directed radially outwards and is symmetrical in all directions (figure 11.1(*a*)).

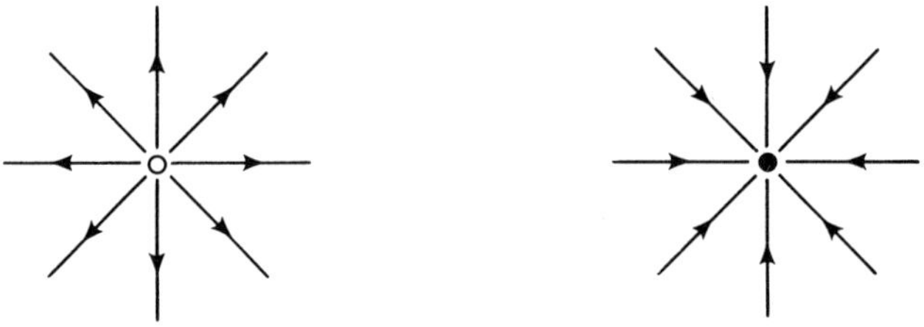

(*a*) (*b*)

Figure 11.1 (*a*) Point source, (*b*) point sink.

A sink, on the other hand, is a negative source. If a source is a point of creation then a sink is a point of annihilation of the fluid. The flow is directed radially inwards (figure 11.1(*b*)) and the potential ϕ for a sink is given by

$$\phi=-m/4\pi r. \tag{11.5}$$

11.2.2 Flow near an infinite plate

The method of images can be used to determine the velocity potential near an infinite plate. Consider a point source of strength m situated at $x_1=-a$, ($x_2=x_3=0$) and the plate is placed at $x_1=0$. The image source of the same strength may be assumed to be located at $x_1=a$, ($x_2=x_3=0$) which is depicted in figure 11.2. The resulting potential ϕ at any point P due to the source

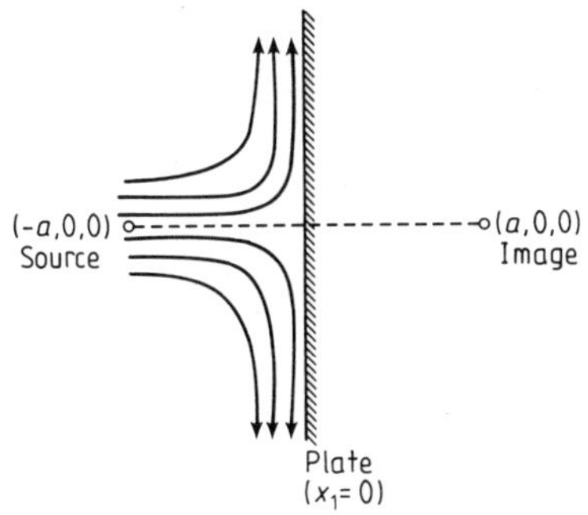

Figure 11.2 Flow near an infinite fixed plate.

and its image can be written as

$$\phi=\frac{m}{4\pi}\left(\frac{1}{[(x_1-a)^2+x_2^2+x_3^2]^{1/2}}+\frac{1}{[(x_1+a)^2+x_2^2+x_3^2]^{1/2}}\right)$$

$$=\frac{m}{4\pi}\left(\frac{1}{r_1}+\frac{1}{r_2}\right). \tag{11.6}$$

The flow velocity at the point P along the x_1-direction therefore becomes

$$v_{x_1}=-\frac{\partial\phi}{\partial x_1}=\frac{m}{r_1^2}\frac{dr_1}{dx_1}+\frac{m}{r_2^2}\frac{dr_2}{dx_2}$$

or

$$v_{x_1}=\frac{m}{4\pi}\left(\frac{x_1-a}{r_1^3}+\frac{x_1+a}{r_2^3}\right). \tag{11.7}$$

If the point P lies at the plate ($x_1=0, r_1=r_2$) then $v_{x_1}=0$, there is no flow along the x_1-direction. This is true for a point P lying anywhere in the plane $x_1=0$. The surface of the plate is therefore a stream surface.

11.3 Liquid streaming past a fixed sphere

11.3.1 Velocity potential

Consider that an incompressible fluid flowing with velocity V ($v_{x_1}=V$; $v_{x_2}=v_{x_3}=0$) along the positive x_1-axis is passing a fixed rigid sphere. The boundary conditions (we shall use spherical polar coordinates r, θ, φ) for the flow past the fixed sphere are:

(i) The flow remains unaffected far from the sphere,

$$\phi=-Vr\cos\theta \tag{11.8a}$$

for $r\to\infty$.

$$v_r=-\frac{\partial\phi}{\partial r}=V\cos\theta \tag{11.8b}$$

(ii) The radial flow velocity v_r at the surface of the sphere is zero,

$$v_r = -\partial\phi/\partial r = 0 \qquad \text{at } r=a \text{ (radius of the sphere).} \tag{11.9}$$

Let us now express Laplace's equation (11.1*a*) in spherical polar coordinates (r, θ, φ):

$$\frac{1}{r^2}\frac{\partial}{\partial r}\left(r^2\frac{\partial\phi}{\partial r}\right)+\frac{1}{r^2 \sin\theta}\frac{\partial}{\partial\theta}\left(\sin\theta\frac{\partial\phi}{\partial\theta}\right)+\frac{1}{r^2\sin^2\theta}\frac{\partial^2\phi}{\partial\varphi^2}=0. \tag{11.10}$$

The solution of equation (11.10) is a standard problem of mathematical physics and can be found in any text book. The general solution has the form

$$\phi=\sum_{n=0}^{\infty}\sum_{m=-n}^{n}\left(a_n r^n+\frac{b_n}{r^{n+1}}\right)Y_{nm}(\theta,\varphi) \tag{11.11}$$

where a_n, b_n are coefficients and $Y_{nm}(\theta, \varphi)$ represent the spherical harmonics,

$$Y_{nm}(\theta,\varphi)=c_{nm}P_n^m(\cos\theta)\exp(\pm im\varphi) \tag{11.12}$$

where c_{nm} are normalisation constants. When equation (11.11) is referred to the boundary condition (11.8*a*) it becomes apparent that the spherical harmonics, $Y_{nm}(\theta, \varphi)$, must reduce to $\cos\theta$. This is possible only when $n=1$ and $m=0$ for which $Y_{10}=c_{10}P_1(\cos\theta)=c_{10}\cos\theta$. Therefore equation (11.11), for our purposes, reduces to

$$\phi=(A_{10}r+B_{10}r^{-2})\cos\theta. \tag{11.13}$$

The coefficients A_{10} $(=a_1c_{10})$ and B_{10} $(=b_1c_{10})$ are determined with the help of the boundary conditions (11.8) and (11.9) respectively,

$$A_{10}=-V \tag{11.14}$$

$$B_{10}=-\tfrac{1}{2}Va^3. \tag{11.15}$$

Substituting these values back into equation (11.13) one obtains the final expression for the velocity potential ϕ:

$$\phi=-V\cos\theta(r+a^3/2r^2). \tag{11.16}$$

Therefore, the flow velocity components become

$$v_r=-\frac{\partial\phi}{\partial r}=V\cos\theta\left(1-\frac{a^3}{r^3}\right) \tag{11.17a}$$

$$v_\theta=-\frac{1}{r}\frac{\partial\phi}{\partial\theta}=-V\sin\theta\left(1+\frac{a^3}{2r^3}\right) \tag{11.17b}$$

$$v_\varphi=\frac{1}{r\sin\theta}\frac{\partial\phi}{\partial\varphi}=0. \tag{11.17c}$$

The flow pattern of liquid past a fixed sphere of radius a is depicted

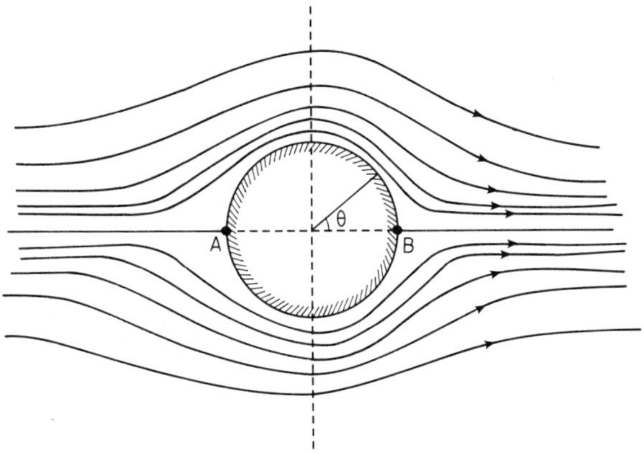

Figure 11.3 Flow past a fixed sphere. A and B are stagnant points.

schematically in figure 11.3. At $\theta=0$ and π, $v_\theta=0$ and hence these are called stagnation points.

11.3.2 Pressure and force on the sphere (d'Alembert's paradox)

In order to calculate the pressure on different parts of the sphere we shall make use of Bernoulli's equation which was deduced earlier. For steady flow ($\partial\phi/\partial t=0$) and in the absence of external forces ($U=0$), Bernoulli's equation is (see equation (10.22))

$$p=\text{constant}-\tfrac{1}{2}\rho \mathbf{v}^2 \tag{11.18}$$

where

$$\mathbf{v}^2=v_r^2+v_\theta^2. \tag{11.19}$$

At the surface of the sphere ($r=a$), equations (11.19) and (11.17) imply that

$$\mathbf{v}^2=\tfrac{9}{4}V^2\sin^2\theta. \tag{11.20}$$

Therefore, equation (11.18) becomes,

$$p=\text{constant}-\tfrac{9}{8}\rho V^2\sin^2\theta. \tag{11.21}$$

This is the required expression for the pressure which acts normally inwards on each point of the sphere. The pressure at stagnation points is a maximum. From the symmetry of equation (11.21) it is obvious that the pressure at any two symmetric points on the sphere is the same in magnitude but the opposite in direction, for instance, pairs of points such as A and B, C and D as shown in figure 11.4. Thus, there is no resultant pressure and hence no net force is acting on the sphere. The latter conclusion becomes clear if we consider the force

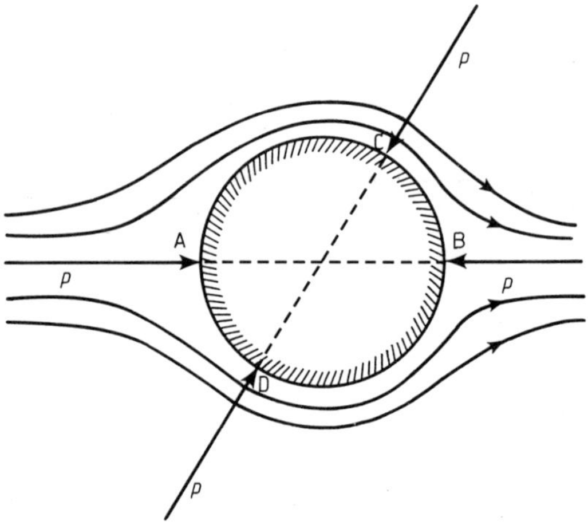

Figure 11.4 Pressure on a fixed sphere inside a streaming liquid.

components (X, Y, Z) at $r=a$:

$$X=-\int_0^{\pi}\int_0^{2\pi} p\cos\theta\,\mathrm{d}A \tag{11.22a}$$

$$Y=-\int_0^{\pi}\int_0^{2\pi} p\sin\theta\cos\varphi\,\mathrm{d}A \tag{11.22b}$$

$$Z=-\int_0^{\pi}\int_0^{2\pi} p\sin\theta\sin\varphi\,\mathrm{d}A \tag{11.22c}$$

where $\mathrm{d}A$ is the elemental surface area of the sphere at the point considered:

$$\mathrm{d}A=a^2\sin\theta\,\mathrm{d}\theta\,\mathrm{d}\varphi. \tag{11.23}$$

It follows that

$$X=Y=Z=0 \tag{11.24}$$

i.e. the net force acting on a fixed sphere in a streaming liquid (inviscid) is zero. This is d'Alembert's paradox.

11.4 A rigid sphere moving through a liquid

11.4.1 Velocity potential

Let the sphere be moving along the x_1-direction with velocity V' in a liquid at rest at $r=\infty$ (r is measured from the centre of the sphere). The boundary conditions are:

(i) At $r=\infty$, the liquid is at rest, therefore:

$$v_r = v_\theta = v_\varphi = 0. \tag{11.25}$$

(ii) At the surface ($r=a$) of the sphere the flow velocity is simply the component of V' along a given r, i.e.

$$v_r = -\partial\phi/\partial r = V' \cos\theta \qquad \text{at } r=a. \tag{11.26}$$

Now the mathematical detail is very similar to the previous example. When the boundary conditions (11.25) and (11.26) are subjected to (11.11), one obtains an expression for the potential ϕ:

$$\phi = (V'a^3/2r^2)\cos\theta. \tag{11.27}$$

The flow velocity components become,

$$v_r = -\frac{\partial\phi}{\partial r} = \frac{V'a^3 \cos\theta}{r^3} \tag{11.28a}$$

$$v_\theta = -\frac{1}{r}\frac{\partial\phi}{\partial\theta} = \frac{V'a^3 \sin\theta}{2r^3} \tag{11.28b}$$

$$v_\varphi = 0. \tag{11.28c}$$

The instantaneous flow pattern is therefore the same as shown in figure 11.3.

11.4.2 Force on the sphere

For uniform motion ($\partial\phi/\partial t = 0$) of the sphere the resultant pressure and hence the force on the sphere would be zero as discussed in §11.3.2. For non-uniform motion ($\partial\phi/\partial t \neq 0$), the resultant force is not zero. The pressure in this case is obtained by employing the generalised form of Bernoulli's equation (see equation (10.21*b*)),

$$p = \rho\frac{\partial\phi}{\partial t} - \tfrac{1}{2}\rho v^2 + \text{constant} \tag{11.29}$$

where

$$\begin{aligned} v^2 &= v_r^2 + v_\theta^2 \\ &= (V'a^3/r^3)^2(1 - \tfrac{3}{4}\sin^2\theta). \end{aligned} \tag{11.30}$$

The substitution of equation (11.29) into (11.22) yields the force components X, Y, Z. Bearing in mind the form of ϕ from (11.27) the integrations in (11.22*b*) and (11.22*c*) vanish, i.e. $Y = Z = 0$. The only non-zero component of the force is,

$$X = -\int_0^\pi\int_0^{2\pi}\left(\rho\frac{\partial\phi}{\partial t} - \tfrac{1}{2}\rho \boldsymbol{v}^2 + \text{constant}\right)\cos\theta \, \mathrm{d}A \tag{11.31}$$

where $\mathrm{d}A = a^2 \sin\theta \, \mathrm{d}\theta \, \mathrm{d}\varphi$. The integral due to the term $(-\tfrac{1}{2}\rho v^2 + \text{constant})$

vanishes, therefore equation (11.31) reduces to

$$X = -\rho \int_0^{\pi} \int_0^{2\pi} \frac{\partial \phi}{\partial t} \cos \theta \, \mathrm{d}A. \tag{11.32}$$

In order to perform the above integration we require $\partial\phi/\partial t$, the derivation of which follows. Let $(x_0, 0, 0)$ denote the position of the sphere at time t, then

$$V' = \dot{x}_0(t) \tag{11.33}$$

$$\cos \theta = (x_1 - x_0)/r \tag{11.34}$$

$$r^2 = (x_1 - x_0)^2 + x_2^2 + x_3^2 = (x_1 - x_0)^2 \tag{11.35}$$

or

$$r\dot{r} = -(x_1 - x_0)V' \tag{11.36}$$

where the dot stands for the first derivative with respect to time. Equation (11.27) may now be re-expressed as,

$$\phi = V'a^3(x_1 - x_0)/2r^3$$

or

$$\frac{\partial \phi}{\partial t} = \frac{\partial}{\partial t}\left(\frac{V'a^3(x_1 - x_0)}{2r^3}\right). \tag{11.37}$$

Differentiating and making use of the different equalities of equations (11.33)–(11.36), one arrives at

$$\left(\frac{\partial \phi}{\partial t}\right)_{r=a} = \tfrac{1}{2}\dot{V}'a \cos \theta + \tfrac{1}{2}(V')^2(3 \cos^2 \theta - 1). \tag{11.38}$$

This is now substituted in (11.32) which, on performing the integration, yields

$$X = -\tfrac{2}{3}\pi\rho a^2 \dot{V}' \tag{11.39}$$

or

$$X = -\tfrac{1}{2}M'\dot{V}' \tag{11.40}$$

where $M' = \frac{4}{3}\pi\rho a^3$ is the mass of liquid displaced by the sphere. It may be noticed that X vanishes when V' is a constant.

11.4.3 Equation of motion

Let the sphere fall with a velocity V' in a liquid under the gravitational force. If ρ is the density of the liquid and σ is the density of the sphere then the gravitational force acting on the sphere is,

$$\tfrac{4}{3}\pi a^3(\sigma - \rho)g = (M - M')g \tag{11.41}$$

where M and M' are the masses of the sphere and the displaced liquid respectively. The equation of motion therefore reads as

$$M\dot{V}' = -\tfrac{1}{2}M'\dot{V}' + (M - M')g$$

or

$$M\dot{V}' = \frac{M - M'}{M + \frac{1}{2}M'} Mg. \tag{11.42}$$

Thus the presence of the liquid reduces the external force of gravity by a factor $(M - M')/(M + \frac{1}{2}M')$.

Problems

11.1 Two point sources of strengths $-m$ and m are located respectively at $(-a, 0, 0)$ and $(a, 0, 0)$.

(i) Show that at a point P far away from the sources the velocity potential is given by,

$$\phi = \frac{ma}{2\pi} \frac{\cos\theta}{r^2} \tag{P11.1}$$

where r is the distance of P from the origin and θ is the angle between $\boldsymbol{r}$ and the x-axis.

(ii) Determine the flow velocity for $r \gg a$ and sketch its variation with θ for a given r.

(iii) If the fluid is incompressible what is the expression for the pressure when ϕ is as in (P11.1)?

11.2 A rigid sphere is moving with a velocity V' in a liquid at rest at infinity. If the flow of the liquid remains irrotational then show that the kinetic energy, T, of the liquid is given by

$$T = \tfrac{1}{4}M'(V')^2 \tag{P11.2}$$

where M' is the mass of liquid displaced by the sphere. (Hint: $T = \int_{\text{vol.}} \frac{1}{2}\rho v^2 \, d^3r$).

12

POTENTIAL FLOW IN TWO DIMENSIONS: COMPLEX-VARIABLE METHOD

If the velocity of a flow is parallel to a fixed plane and is the same in all planes parallel to it then the flow is called two dimensional. For completeness if we consider such a plane as the xy-plane in a Cartesian system† then $v_z=0$ and also $\partial v_x/\partial z=\partial v_y/\partial z=0$. The absence of one space coordinate simplifies the basic equations (such as the incompressibility condition, the Laplace equation, etc) considerably. We shall see that the condition of incompressibility in two dimensions leads us to introduce the stream function ψ which proves very useful in the discussion of flow problems. The function ψ is also closely connected to the velocity potential ϕ through the Laplace equation. This paves the way to an elegant use of the complex-variable method to solve two-dimensional flow problems.

12.1 Stream function

12.1.1 Definition of stream function

The equation of continuity for an incompressible fluid, div $\boldsymbol{v}=0$, in two dimensions is

$$\frac{\partial v_x}{\partial x}+\frac{\partial v_y}{\partial y}=0 \qquad \text{or } \frac{\partial v_y}{\partial y}=-\left(\frac{\partial v_x}{\partial x}\right) \tag{12.1}$$

where v_x and v_y are the velocity components along the x- and y-directions respectively. Because of equation (12.1) one infers that $v_y\,\mathrm{d}x-v_x\,\mathrm{d}y$ is a perfect differential, say $\mathrm{d}\psi$, i.e.

$$\mathrm{d}\psi=v_y\,\mathrm{d}x-v_x\,\mathrm{d}y=\frac{\partial\psi}{\partial x}\,\mathrm{d}x+\frac{\partial\psi}{\partial y}\,\mathrm{d}y \tag{12.2}$$

where

$$v_y=\frac{\partial\psi}{\partial x} \qquad v_x=-\frac{\partial\psi}{\partial y}. \tag{12.3}$$

† In this Chapter we choose $x\equiv x_1$, $y\equiv x_2$, $z\equiv x_3$. This makes the discussion more plausible.

The function,

$$\psi = \psi(x, y) = \int (v_y \, dx - v_x \, dy) \tag{12.4}$$

is called the stream function. The direction of the flow is given by the differential equation,

$$\frac{dx}{v_x} = \frac{dy}{v_y}. \tag{12.5}$$

Multiplying both sides by $v_x v_y$ one obtains

$$v_y \, dx - v_x \, dy = 0$$

or

$$d\psi = 0. \tag{12.6}$$

The integration of equation (12.6) readily yields

$$\psi = \text{constant}.$$

A line joining all the points in the xy plane with $\psi = \text{constant}$, is called a streamline. Corresponding to different values of the constant one obtains a set of such streamlines.

12.1.2 Connection between ψ (stream function) and ϕ (velocity potential)

If the flow is irrotational, curl $\mathbf{v} = 0$, i.e. in two dimensions, one has

$$\frac{\partial v_y}{\partial x} - \frac{\partial v_x}{\partial y} = 0. \tag{12.7}$$

Substituting for v_x and v_y from equation (12.3) it becomes

$$\frac{\partial^2 \psi}{\partial x^2} + \frac{\partial^2 \psi}{\partial y^2} = 0$$

or

$$\nabla^2 \psi = 0. \tag{12.8}$$

Thus ψ satisfies a Laplace equation similar to $\nabla^2 \phi = 0$. In two dimensions $\nabla^2 \phi = 0$ can be expressed as

$$\frac{\partial^2 \phi}{\partial x^2} + \frac{\partial^2 \phi}{\partial y^2} = 0. \tag{12.9}$$

The velocity components are:

$$v_x = -\partial \phi / \partial x \qquad v_y = -\partial \phi / \partial y. \tag{12.10}$$

Equations (12.10) and (12.3) together imply:

$$\frac{\partial \psi}{\partial y}=\frac{\partial \phi}{\partial x} \qquad \frac{\partial \psi}{\partial x}=\frac{-\partial \phi}{\partial y}. \tag{12.11}$$

These are known as the Cauchy–Riemann equations. From their very existence we may also conclude that

$$\frac{\partial \phi}{\partial x}\frac{\partial \psi}{\partial x}+\frac{\partial \phi}{\partial y}\frac{\partial \psi}{\partial y}=0 \tag{12.12}$$

which shows that the curves, ϕ = constant (equipotential lines) and ψ = constant (streamlines) always intersect at right angles. Thus equipotential lines and streamlines are orthogonal to one another (figure 12.1). It is also important to notice from the Cauchy–Riemann equation (12.11) that ϕ and ψ are conjugate functions and therefore can be expressed as a single complex-variable function called the complex potential. This paves the way to use the theory of complex-variable functions to solve two-dimensional flow problems.

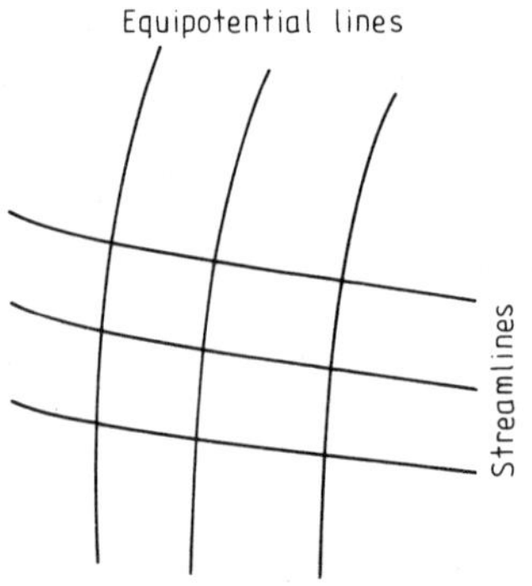

Figure 12.1 Orthogonal network of streamlines and equipotential lines.

12.2 Complex potential Ω and its relation to ϕ and ψ

We shall see that the Cauchy–Riemann equations follow if we assume the existence of a complex potential $\Omega(z)$ (z is a complex number, $z=x+\mathrm{i}y$, i stands for $\sqrt{-1}$) such that,

$$\begin{aligned}\Omega(z)&=\Omega(x+\mathrm{i}y)\\ &=\phi(x, y)+\mathrm{i}\psi(x, y)\equiv\phi+\mathrm{i}\psi.\end{aligned} \tag{12.13}$$

We therefore write

$$\frac{\partial \Omega}{\partial x}=\frac{\mathrm{d}\Omega}{\mathrm{d}z}\frac{\partial z}{\partial x}=\frac{\mathrm{d}\Omega}{\mathrm{d}z} \tag{12.14}$$

$$\frac{\partial\Omega}{\partial y}=\frac{\mathrm{d}\Omega}{\mathrm{d}z}\frac{\partial z}{\partial y}=\mathrm{i}\frac{\mathrm{d}\Omega}{\mathrm{d}z}. \tag{12.15}$$

Thus

$$\mathrm{i}\frac{\partial\Omega}{\partial x}=\frac{\partial\Omega}{\partial y} \tag{12.16}$$

or

$$\mathrm{i}\left(\frac{\partial\phi}{\partial x}+\mathrm{i}\frac{\partial\psi}{\partial x}\right)=\frac{\partial\phi}{\partial y}+\mathrm{i}\frac{\partial\psi}{\partial y}. \tag{12.17}$$

By equating the real and imaginary parts one recovers the Cauchy–Riemann differential equation,

$$\partial\psi/\partial y=\partial\phi/\partial x \qquad \partial\psi/\partial x=-\partial\phi/\partial y$$

where ϕ and ψ are conjugate functions. Thus, two-dimensional flow problems can be represented by an analytic function $\Omega(z)$ whose real part gives the velocity potential ϕ and the imaginary part gives the stream function ψ,

$$\phi=\mathrm{Re}\,\Omega(z) \qquad \psi=\mathrm{Im}\,\Omega(z). \tag{12.18}$$

At this stage we wish to emphasise that the identification of ϕ and ψ as conjugate functions in the theory of functions of a complex variable is very important. This provides an advantage in using the method of conformal mapping to solve complicated flow problems. If the flow around an object of simple geometrical shape is known then the flow pattern around more complex shapes could be inferred by conformal mapping. We do not propose to pursue this problem any further, but we intend to limit our discussion to some simple forms of $\Omega(z)$. (Fuller accounts may be found in many other text books such as Milne-Thomson (1960).)

Given the complex potential $\Omega(z)$ one can easily determine the complex velocity and the flow velocity. From equations (12.13) and (12.14) we have

$$\frac{\mathrm{d}\Omega}{\mathrm{d}z}=\frac{\partial\Omega}{\partial x}$$

$$=\frac{\partial\phi}{\partial x}+\mathrm{i}\frac{\partial\psi}{\partial x}$$

since

$$\partial\phi/\partial x=-v_x \qquad \text{and } \partial\psi/\partial x=v_y$$

therefore

$$\mathrm{d}\Omega/\mathrm{d}z=-(v_x-\mathrm{i}v_y)=-\upsilon. \tag{12.19}$$

$\upsilon=v_x-\mathrm{i}v_y$ is called the complex velocity. The multiplication of equation

(12.19) by its complex conjugate enables us to write,

$$\left|\frac{\mathrm{d}\Omega}{\mathrm{d}z}\right|^2 = |\mathcal{v}|^2 = v_x^2 + v_y^2$$

or

$$|\mathcal{v}| = v = (v_x^2 + v_y^2)^{1/2} \tag{12.20}$$

where v is the flow velocity.

12.3 Some simple forms of $\Omega(z)$

First consider,

$$\Omega(z) = Az^n \tag{12.21}$$

where $z = x + \mathrm{i}y$ and A is a real constant. The exponent n may either be a positive or a negative number. In polar coordinates (r, φ), z may be expressed in the form,

$$z = r\,\mathrm{e}^{\mathrm{i}\varphi} = r(\cos\varphi + \mathrm{i}\sin\varphi) \tag{12.22}$$

where r is usually called the modulus of z and φ is called the argument of z. Therefore, the complex potential $\Omega(z)$ may now be written as

$$\begin{aligned}\Omega(z) &= Ar^n\,\mathrm{e}^{in\varphi}\\ &= Ar^n(\cos n\varphi + \mathrm{i}\sin n\varphi).\end{aligned} \tag{12.23}$$

By making comparison of (12.23) and (12.13) we infer

$$\phi = Ar^n \cos n\varphi \tag{12.24a}$$

$$\psi = Ar^n \sin n\varphi \tag{12.24b}$$

Readers may verify that (12.24) represents the solution of Laplace's equation. For further discussion we now consider definite values of n:

(i) $n = 1$

Equation (12.21) reduces to

$$\Omega(z) = Az \tag{12.25}$$

or

$$\mathrm{d}\Omega/\mathrm{d}z = A = -\mathcal{v}. \tag{12.26}$$

The last equality in equation (12.26) follows from (12.19). By substituting the values of A and n in (12.24) we obtain

$$\phi = -\mathcal{v}r\cos\varphi = -vx \tag{12.27}$$

$$\psi = -\mathcal{v}r\sin\varphi = -vy. \tag{12.28}$$

Therefore,

$$v_x = -\partial\phi/\partial x = -\partial\psi/\partial y = v \tag{12.29}$$

$$v_y = -\partial\phi/\partial y = \partial\psi/\partial x = 0. \tag{12.30}$$

Thus the complex potential (12.25) corresponds to uniform flow along the x-direction with velocity $|v| = v$. The equipotential lines ϕ = constant, i.e. from equation (12.27) x = constant, are a system of straight lines parallel to the y-axis and the streamlines ψ = constant, i.e. from equation (12.28) y = constant, are a system of straight lines parallel to the x-axis (see figure 12.2).

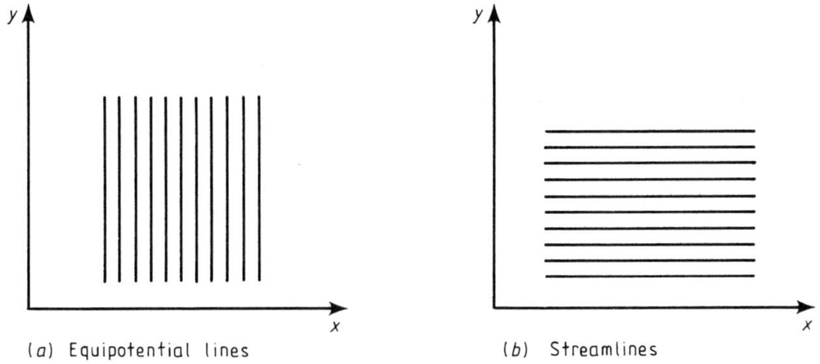

Figure 12.2 Flow pattern for the complex potential $\Omega(z) = Az$.

(ii) $n = 2$

The complex potential $\Omega(z)$ becomes,

$$\Omega(z) = Az^2. \tag{12.31}$$

By employing equation (12.24), ϕ and ψ can be expressed as

$$\phi = Ar^2 \cos 2\varphi = Ar^2(\cos^2 \varphi - \sin^2 \varphi) \tag{12.32}$$

$$\psi = Ar^2 \sin 2\varphi = 2Ar^2 \sin \varphi \cos \varphi. \tag{12.33}$$

Since $x = r \cos \varphi$ and $y = r \sin \varphi$,

$$\phi = A(x^2 - y^2) \tag{12.34}$$

$$\psi = 2Axy. \tag{12.35}$$

The equipotential lines ϕ = constant, i.e. $x^2 - y^2$ = constant represent a series of equilateral hyperbolas symmetrical around the x-axis. Similarly, the streamlines, ψ = constant, i.e. xy = constant, represent a series of equilateral hyperbolas asymptotic to the coordinate axes. The streamlines round a

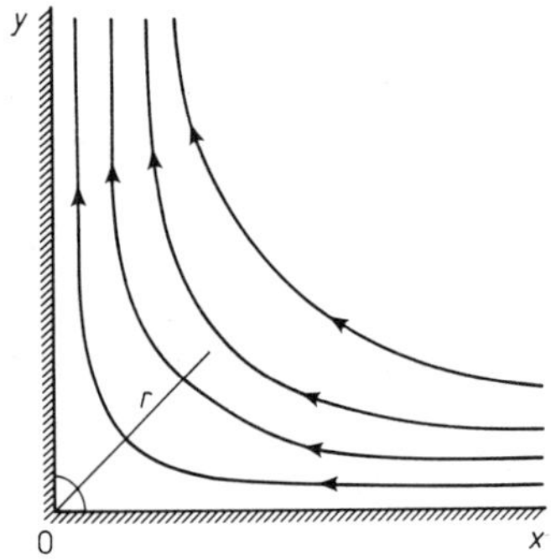

Figure 12.3 Streamlines round a rectangular wall.

rectangular rigid wall are shown in figure 12.3. The velocity of flow is given by

$$v=\left|\frac{\mathrm{d}\Omega}{\mathrm{d}z}\right|=2A|z|=2Ar. \tag{12.36}$$

At the corner of the rectangular wall, i.e. at $r=0$, $v=0$.

12.4 Line source

Consider a logarithmic function for $\Omega(z)$,

$$\Omega(z)=-A\ln z. \tag{12.37}$$

With $z=r\,\mathrm{e}^{\mathrm{i}\varphi}$, equation (12.37) becomes

$$\Omega(z)=-(A\ln r+\mathrm{i}A\varphi). \tag{12.38}$$

Therefore, it follows that

$$\phi=-A\ln r \tag{12.39}$$

$$\psi=-A\varphi. \tag{12.40}$$

The equipotential lines, $\phi=$constant, i.e. $r=$constant, are circles and streamlines, $\psi=$constant, i.e. $\varphi=$constant, are straight radial lines with a common centre at the origin O. These are shown in figure 12.4. The radial and angular velocity, therefore, become

$$v_r=-\partial\phi/\partial r=A/r \tag{12.41}$$

$$v_\varphi=0. \tag{12.42}$$

It may be noticed that $v_r\to\infty$ as $r\to 0$, i.e. o is a singular point.

Let us now consider the amount of fluid flowing per unit time across a circle

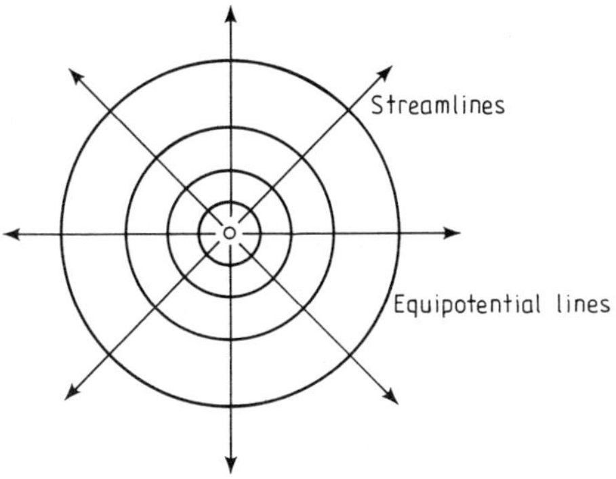

Figure 12.4 Flow pattern from a line source.

of radius r around o (per unit length in the direction perpendicular to the xy-plane),

$$m = 2\pi r v_r = (2\pi r)A/r = 2\pi A \tag{12.43}$$

where m is called the strength of the line source. Equation (12.43) also determines the value of the constant $A = m/2\pi$, and thus

$$\Omega(z) = -(m/2\pi)\ \ln z. \tag{12.44}$$

This represents a line source of strength m. A source with strength $-m$ is called a sink.

12.5 On vortices

If in the line source problem the role of ϕ and ψ is interchanged then equipotentials become radial lines and streamlines become circles. It means that the fluid circulates round and round about the point of origin O as shown in figure 12.5. The potential $\Omega(z)$ for such a flow is represented by

$$\Omega(z) = (\mathrm{i}\kappa/2\pi)\ \ln z \tag{12.45}$$

where κ is real and is called the vortex strength. With $z = r\ \mathrm{e}^{\mathrm{i}\varphi}$, equation (12.45) becomes

$$\Omega(z) = -\frac{\mathrm{i}\kappa}{2\pi}\ \varphi + \frac{\mathrm{i}\kappa}{2\pi}\ \ln r. \tag{12.46}$$

Therefore, the functions ϕ and ψ can be expressed as

$$\phi = -\kappa\varphi/2\pi \qquad \psi = (\kappa/2\pi)\ \ln r. \tag{12.47}$$

Note that (12.47) also satisfies Laplace's equation. The velocity components in

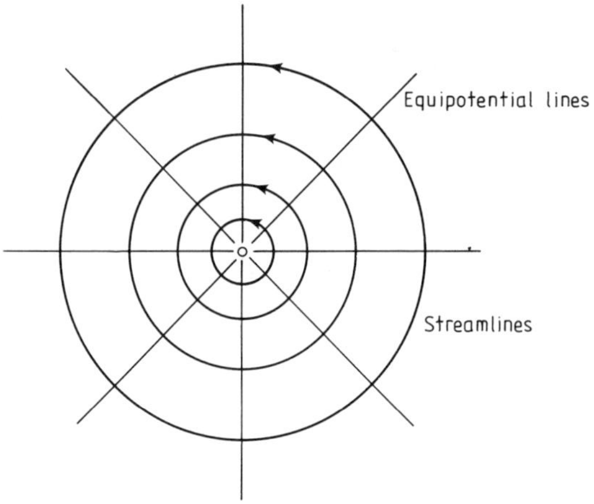

Figure 12.5 Circulation round a vortex O.

the present case are given by

$$v_r = -\partial\phi/\partial r = 0 \tag{12.48a}$$

$$v_\varphi = -\frac{1}{r}\frac{\partial\phi}{\partial\varphi} = \frac{\kappa}{2\pi r} \tag{12.48b}$$

which allows us to write

$$v = \left|\frac{\mathrm{d}\Omega}{\mathrm{d}z}\right| = (v_r^2 + v_\varphi^2)^{1/2} = \frac{\kappa}{2\pi r} \tag{12.49a}$$

or

$$\kappa = 2\pi v r. \tag{12.49b}$$

Thus the vortex strength κ is defined as the flow velocity times the circumference of the unit circle around the origin.

The circulation, Γ, around a circle of radius r encircling the origin O (or, of course, any closed curve which includes the origin) may, therefore, be written as

$$\Gamma = \int v \cdot \mathrm{d}s = \int_0^{2\pi} \left(\frac{\kappa}{2\pi r}\right) r \, \mathrm{d}\varphi$$

or

$$\Gamma = \kappa. \tag{12.50}$$

This shows that the circulation for any closed curve is the same and is equal to the vortex strength. κ is positive when the circulation is counted counterclockwise and is negative otherwise. It is important to observe that for any closed curve the circulation does not vanish but the vorticity ω is always zero[†] except at the origin. We may, therefore, conclude that the flow is a potential (irrotational) flow.

The complex potential $\Omega(z)$ and the circulation Γ corresponding to several vortices located in the z-plane at $z=-a_n$ $(n=1, 2, \ldots)$ can be expressed as,

$$\Omega(z)=\sum_n \frac{\mathrm{i}\kappa_n}{\pi} \ln (z-a_n) \tag{12.51a}$$

$$\Gamma=\sum_n \kappa_n. \tag{12.51b}$$

We wish to end our discussion on vorticity by drawing an interesting analogy between hydrodynamics and electrodynamics. We start with Ampère's law, i.e. a relationship between the current i and the magnetic field $\boldsymbol{B}$,

$$\oint \boldsymbol{B}\cdot \mathrm{d}\boldsymbol{s}=\mu_0 i \tag{12.52}$$

where μ_0 is called the permeability constant. Equation (12.52) can also be expressed as,

$$\int_A \operatorname{curl} \boldsymbol{B}\, \mathrm{d}A=\int_A \mu_0 \boldsymbol{J}\, \mathrm{d}A$$

or

$$\operatorname{curl} \boldsymbol{B}=\mu_0 \boldsymbol{J} \tag{12.53}$$

where $\boldsymbol{J}$ stands for the current density. From the comparison of expressions (12.52) and (12.53) with those of hydrodynamic expressions, such as $\Gamma=\kappa=\int \boldsymbol{v}\cdot \mathrm{d}\boldsymbol{s}$ and $\omega=\frac{1}{2}\operatorname{curl} \boldsymbol{v}$; one may infer that $\boldsymbol{B}$ corresponds to $\boldsymbol{v}$, the current i corresponds to the vortex strength κ, and the current density $\boldsymbol{J}$ corresponds to the vorticity ω. Furthermore, the law of Biot and Savart which computes $\boldsymbol{B}$ at any point P (see figure 12.6) due to an arbitrary distribution of current i along a curved path,

$$\mathrm{d}\boldsymbol{B}=\frac{\mu_0 i}{4\pi} \frac{\mathrm{d}\boldsymbol{s}\times \boldsymbol{r}}{r^3} \tag{12.54}$$

can be used to determine the velocity due to a vortex line of any shape,

$$\mathrm{d}\boldsymbol{v}=\frac{\kappa}{4\pi} \frac{\mathrm{d}\boldsymbol{s}\times \boldsymbol{r}}{r^3}. \tag{12.55}$$

This analogy has proved very useful in the development of both fields.

[†] Readers should verify this by computing the vorticity $\omega=\frac{1}{2}\operatorname{curl} \boldsymbol{v}$.

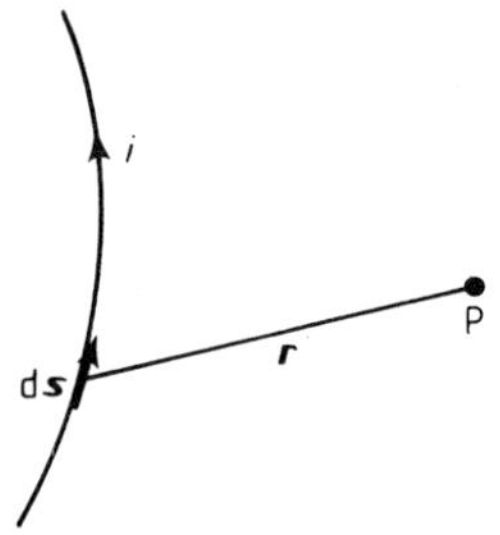

Figure 12.6 Magnetic field at a point P due to current i along a curved path.

12.6 Flow past a long circular cylinder with vortices

12.6.1 Complex potential and stagnation points

Consider a flow in the xy-plane which passes a long circular cylinder. The axis of the cylinder is taken parallel to the z-axis. At a large distance from the cylinder the fluid flows with undisrupted velocity V parallel to the x-axis. If a is the radius of a circular cross section of a cylinder whose centre is at the origin then the obvious boundary conditions are:

(i)

$$v_r = V \cos \varphi \qquad \text{for } r \to \infty. \tag{12.56}$$

(ii) At the surface $r=a$, the flow must be a streamline, i.e.

$$v_r = 0 \qquad \text{at } r = a. \tag{12.57}$$

We shall see that these boundary conditions are satisfied immediately if we take $\Omega(z)$ to be of the form,

$$\Omega(z) = -V(z + a^2/z) + (i\kappa/2\pi) \ln z. \tag{12.58}$$

Working this out in polar coordinates, $z = r\,e^{i\varphi}$, equation (10.58) allows us to express the functions ϕ and ψ as

$$\phi = -V \cos \varphi(r + a^2/r) - \kappa\varphi/2\pi \tag{12.59}$$

$$\psi = -V \sin \varphi(r - a^2/r) + (\kappa/2\pi) \ln r. \tag{12.60}$$

This readily satisfies the boundary conditions (12.56) and (12.57). The velocity components become:

$$v_r = -\partial\phi/\partial r = -V \cos \varphi(1 - a^2/r) \tag{12.61a}$$

$$v_\varphi = -r^{-1}\, \partial\phi/\partial\varphi = -V \sin \varphi(1 + a^2/r^2) + \kappa/2\pi r. \tag{12.61b}$$

At the surface of the cylinder, $r = a$,

$$v_r = 0 \tag{12.62a}$$

$$v_\varphi = -2V \sin \varphi + \kappa/2\pi a. \tag{12.62b}$$

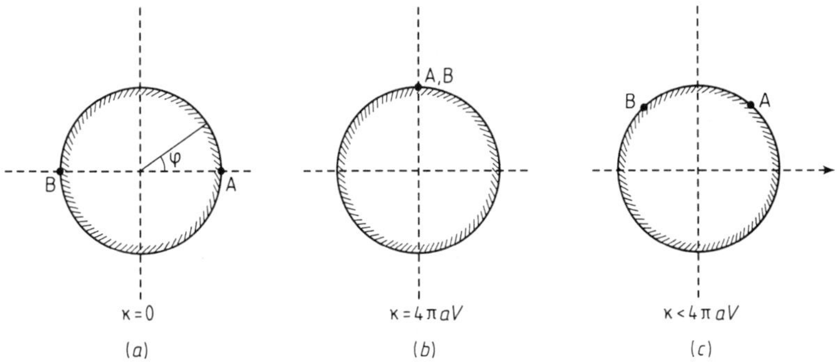

Figure 12.7 Stagnation points A and B.

Thus the points of zero velocity (these are also called stagnation points) at the surface of the cylinder are given by

$$\sin \varphi = \kappa/4\pi a V. \tag{12.63}$$

It is evident that the position of the stagnation points depends upon the vortex strength κ. For $\kappa=0$, the stagnation points are $\varphi=0$ and $\varphi=\pi$, the diametrically opposite points along the x-axis. These are shown in figure 12.7(*a*) as the points A and B. At B, the incoming stream divides and then joins together again at A.

For $\kappa=4\pi a V$ the two stagnation points coincide at $\varphi=\pi/2$, shown in figure 12.7(*b*).

For $\kappa<4\pi a V$ the stagnation points A and B lie in the upper hemisphere, as shown in figure 12.7(*c*). For $\kappa>4\pi a V$, however, equation (12.63) is no longer satisfied and therefore the stagnation points do not lie at the surface of the cylinder.

12.6.2 Force on a cylinder

Recalling Bernoulli's theorem, the pressure p on a cylinder is given by

$$p = \text{constant} - \tfrac{1}{2}\rho v^2$$

where the flow velocity v is

$$\begin{aligned} v^2 &= v_r^2 + v_\varphi^2 \\ &= [(\kappa/2\pi a) - 2V \sin \varphi]^2. \end{aligned} \tag{12.64}$$

Therefore,

$$p = \text{constant} - \tfrac{1}{2}\rho[(\kappa/2\pi a) - 2V \sin \varphi]^2. \tag{12.65}$$

Let X and Y denote the x- and y-components of the force acting per unit length of the cylinder,

$$X=-\int p\cos\varphi\,\mathrm{d}A \tag{12.66}$$

$$Y=-\int p\sin\varphi\,\mathrm{d}A \tag{12.67}$$

where $\mathrm{d}A=a\,\mathrm{d}\varphi\,\mathrm{d}z$. In view of (12.65), these equations may be re-expressed as,

$$\begin{aligned}X&=-\int_0^1 \mathrm{d}z\int_0^{2\pi} a\cos\varphi\,\mathrm{d}\varphi\{\text{constant}-\tfrac{1}{2}\rho[(\kappa/2\pi a)-2V\sin\varphi]^2\}\\&=0\end{aligned} \tag{12.68}$$

and

$$\begin{aligned}Y&=-\int_0^1 \mathrm{d}z\int_0^{2\pi} a\sin\varphi\,\mathrm{d}\varphi\{\text{constant}-\tfrac{1}{2}\rho[(\kappa/2\pi a)-2V\sin\varphi]^2\}\\&=-\frac{\rho V\kappa}{\pi}\int_0^{2\pi}\sin^2\varphi\,\mathrm{d}\varphi\\&=-\rho\kappa V.\end{aligned} \tag{12.69}$$

Thus there is a force on the cylinder in the y-direction which is proportional to the fluid velocity V and the vortex strength κ. Some interesting observations follow:

(i) In the absence of circulation ($\kappa=0$) one has $X=Y=0$, i.e. there is no net force on the cylinder.

(ii) From the expression (12.69) it is instructive to note that the force on the cylinder acts along the negative y-direction, i.e. downwards. This can also be checked intuitively by examining the flow pattern as shown in figure 12.8. In

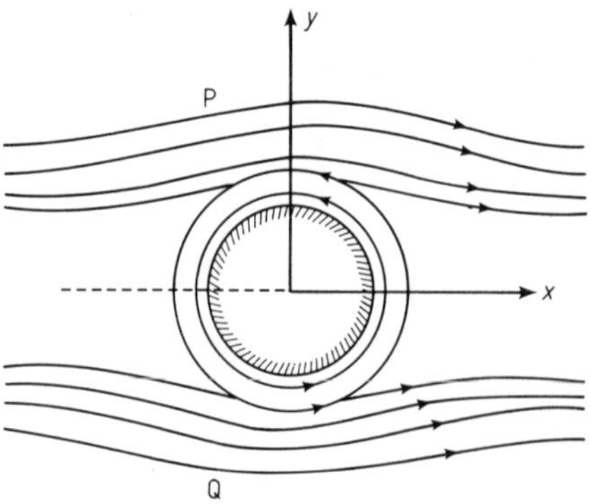

Figure 12.8 Flow past a circular cylinder with circulation.

the upper hemisphere (P) the circulation and the stream velocity are oppositely dirrected, unlike the lower hemisphere (Q). This means that the resultant velocity at P is smaller than at Q, i.e. $v_P^2 < v_Q^2$. By virtue of Bernoulli's theorem it is therefore suggested that $p_P > p_Q$, hence the pressure acts downwards.

(iii) For $V = -V$, i.e. if the fluid flows in the negative x-direction then the force, $Y = \rho\kappa V$, acts along the positive y-direction. This time the circulation and the stream velocity V are oppositely directed in the lower hemisphere and therefore the thrust acts upwards. Such a force is usually called the lift because it has the tendency to lift the cylinder upwards. We may conclude that the circulation together with the stream flow (parallel to the x-axis) cause a lift. Though our conclusion is based on a result for an infinitely long circular cylinder, it is true for many other shapes. The principle of lift is also known as the theorem of Kutta and Joukowski.

(iv) With a view to forming some ideas regarding the lift, we consider a simple numerical example. Air flows past a circular cylinder of radius a. If $v \sim V$, then the vortex strength, $\kappa = \oint \boldsymbol{v} \cdot d\boldsymbol{s} \simeq 2\pi a v$ and the force Y becomes

$$Y = 2\pi a \rho_{\text{air}} v^2. \tag{12.70}$$

On the other hand, if ρ_c is the density of the material of the cylinder then the gravitational pull per unit length of the cylinder is $\pi a^2 \rho_c g$. The uplift is

$$2\pi a \rho_{\text{air}} v^2 = \pi a^2 \rho_c g$$

or

$$v = (a g \rho_c / 2\rho_{\text{air}})^{1/2}. \tag{12.71}$$

With $\rho_{\text{air}} \simeq 1.5 \times 10^{-3}$ g cm^{-3}, $\rho_c \simeq 3$ g cm^{-3}, $g = 980$ cm s^{-2} and for $a = 1$ cm, one has

$$v \simeq 36 \text{ km h}^{-1}$$

which shows the order of magnitude of the flow velocity necessary to lift such a cylindrical object.

(v) Finally, we wish to emphasise the cause which gives rise to the circulation. When there was initially no rotation in the fluid at any point, how did circulations start round the cylinder? Actually, they arise only if the fluid possesses viscosity. The presence of viscosity causes friction in the boundary layer and introduces a shearing stress which leads to vortices. These features of the fluid shall be considered in the following Chapters.

Problems

12.1 For a flow associated with a line vortex the velocity potential is given by

$$\phi = -A \tan^{-1} y/x \tag{P12.1}$$

with $v_z \equiv 0$.

(i) Determine v_x and v_y and sketch how the particles are moving in the xy-plane.

(ii) The fluid is incompressible and located in the gravitational field which is along the z-axis. Show that the pressure obeys the equation,

$$\frac{p}{\rho}+\frac{A^2}{2r^2}+gz=c \tag{P12.2}$$

where c is a constant and $r^2=x^2+y^2$.

(iii) Evaluate the vorticity $\boldsymbol{\omega}=\frac{1}{2}$ curl $\boldsymbol{v}$.

12.2 Two line sources of equal strength m are located at $(b, 0)$ and $(c, 0)$. A third line source of strength $(-m)$ is located at the origin $(0, 0)$. Determine the expressions for ϕ and ψ and show that the circle,

$$x^2+y^2=bc \tag{P12.3}$$

is one of the streamlines.

12.3 Consider a solution of a liquid streaming past a long circular cylinder (see §12.7.1) with vortex strength $\kappa<4\pi aV$. Sketch the streamlines and compare them with the case $\kappa=0$.

12.4 A circular vortex ring of radius a and strength κ lies in the xy-plane with its centre at the origin of the coordinate axis. Show that the fluid velocity at a point $(0, 0, z)$ on the axis of the ring is given by

$$v=\kappa a^2/2(a^2+z^2)^{3/2}. \tag{P12.4}$$

What is the direction of the velocity?

13

FLUIDS WITH VISCOSITY: THE NAVIER–STOKES EQUATION

In the preceding Chapters on fluids we entirely confined ourselves to perfect (inviscid) fluids. The remaining Chapters will be devoted to the problems of real fluids (i.e. fluids with viscosity). We saw earlier that the stress on an element of surface in a perfect fluid is essentially hydrostatic pressure. The stress components are always normal to the surface. This is, however, strictly true only if the fluid is either at rest or in uniform motion. Whenever there is a velocity gradient, the viscous forces become operative. The stress on a surface element then contains both the normal and the shear components.

The stress tensor due to viscosity will be introduced shortly by using symmetry considerations (see Chapter 4) about elasticity†. This is then utilised to obtain the equations of motion, i.e. the Navier–Stokes equation. The viscous forces are by nature dissipative forces. They entail the irreversible conversion of mechanical energy into thermal energy which has been briefly discussed.

13.1 Viscosity stress tensor

13.1.1 Newton's law of viscosity

Consider a fluid moving along the x_1-direction in the x_1x_2-plane with velocity v_1. (v_1 is supposed to be a function of x_2 only), and the velocity gradient $\mathrm{d}v_1/\mathrm{d}x_2>0$. Let us choose a reference layer of the fluid at $x_2=0$ (see figure 13.1); the layer below moves slowly while the layer above moves faster than the layer at $x_2=0$. In other words, taking a microscopic view, the molecules in the layer $x_2<0$ are moving slower, on the average, than the molecules at $x_2>0$. This causes an exchange of molecules or a transfer of momentum between the fluid layers. In essence, the lower layer exerts a drag force on the $x_2=0$ layer tending to slow it down while the upper layer tends to accelerate it. A similar situation exists for each layer. The force obviously acts in a plane perpendicular to the x_2-axis and is directed along the x_1-axis. Following the definition of stress components (see §4.1) let us denote this force per unit area

† The two phenomena, elasticity and viscosity are physically quite different but the mathematical approach is very similar.

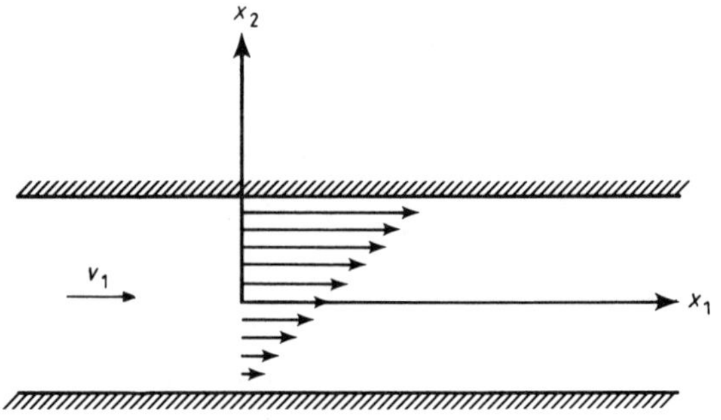

Figure 13.1 Velocity gradient between different layers in a viscous fluid.

by the stress component σ_{21}. If $\mathrm{d}v_1/\mathrm{d}x_2$ is small then it is found that

$$\sigma_{21}=\eta(\mathrm{d}v_1/\mathrm{d}x_2). \tag{13.1}$$

Equation (13.1) is often referred to as Newton's law of viscosity. The constant of proportionality, η, is called the shear viscosity or the first coefficient of viscosity. If L, M and T represent the units of length, mass and time then obviously $\mathrm{ML}^{-1}\mathrm{T}^{-1}$ are the dimensions of η. In the CGS system it is measured in grams per centimetre per second or simply poise.

13.1.2 Generalisation of equation (13.1)

If all the $\partial v_i/\partial x_j$ vanish everywhere (the fluid is at rest or in uniform motion) then

$$\sigma_{11}=\sigma_{22}=\sigma_{33}=-p$$

$$\sigma_{ij}=0 \qquad \text{for } i\neq j.$$

We therefore generalise the stress components in the form,

$$\sigma_{ij}=-p\delta_{ij}+\Pi_{ij} \qquad i,j=1,2,3 \tag{13.2}$$

where Π_{ij} represent the viscosity stress tensor and are identically zero if all $\partial v_i/\partial x_j=0$. Like σ_{ij}, these are also components of a symmetric tensor of rank two.

We now assume that $\partial v_i/\partial x_j$ are sufficiently small and therefore Π_{ij} may be expressed as a linear function of $\partial v_i/\partial x_j$. In order to establish the relationship we shall work in close analogy with the symmetry considerations given in Chapters 3 and 4. Similar to the strain tensor ε_{ij} (equation (3.12)), we first introduce the velocity gradient tensor v_{ij}[†],

$$v_{ij}=\frac{1}{2}\left(\frac{\partial v_i}{\partial x_j}+\frac{\partial v_j}{\partial x_i}\right). \tag{13.3}$$

[†] This is also known as the rate of strain tensor ($v_{ij}=\partial\varepsilon_{ij}/\partial t$).

Secondly, fluids are isotropic therefore analogous to the stress–strain relation for an elastic isotropic solid ($\sigma_{ij}=\lambda(\varepsilon_{11}+\varepsilon_{22}+\varepsilon_{33})\delta_{ij}+2\mu\varepsilon_{ij}$, equation (4.65)), we express the relationship between Π_{ij} and v_{ij} as

$$\Pi_{ij}=\eta'(v_{11}+v_{22}+v_{33})\delta_{ij}+2\eta v_{ij} \tag{13.4a}$$

$$=\eta' \operatorname{div} \boldsymbol{v}\delta_{ij}+2\eta v_{ij}. \tag{13.4b}$$

Here η and η' are known as the first and second coefficients of viscosity and take the place of Lame's constants μ ($\equiv c_{44}$) and λ ($\equiv c_{12}$) respectively.

By substituting equation (13.4*b*) into (13.2), we obtain the stress components for a real fluid:

$$\sigma_{ij}=-p\delta_{ij}+\eta' \operatorname{div} \boldsymbol{v}\delta_{ij}+2\eta v_{ij}. \tag{13.5}$$

For an incompressible fluid, div $\boldsymbol{v}=0$, equation (13.5) reduces to

$$\sigma_{ij}=-p\delta_{ij}+2\eta v_{ij}. \tag{13.6}$$

13.2 Bulk and shear viscosity

Let us define, as usual, the mean pressure $\bar{p}$ as

$$\bar{p}=-\tfrac{1}{3}(\sigma_{11}+\sigma_{22}+\sigma_{33}).$$

Making a substitution for σ_{ii} from equation (13.5) the mean pressure for a real fluid becomes

$$\bar{p}=p-(\eta'+\tfrac{2}{3}\eta) \operatorname{div} \boldsymbol{v}. \tag{13.7}$$

Remembering the equation of continuity ($\mathrm{D}\rho/\mathrm{D}t+\rho_0 \operatorname{div} \boldsymbol{v}=0$), the above relation can also be expressed as

$$\bar{p}=p+\frac{(\eta'+\tfrac{2}{3}\eta)}{\rho_0}\frac{\mathrm{D}\rho}{\mathrm{D}t}$$

$$=p+\frac{b}{\rho_0}\frac{\mathrm{D}\rho}{\mathrm{D}t} \tag{13.8}$$

where

$$b=\eta'+\tfrac{2}{3}\eta \tag{13.9}$$

is called the coefficient of bulk viscosity. We observe that η and η' are related to the bulk viscosity b through an expression similar to the relation between Lame's constants μ and λ and the bulk modulus B (i.e. $B=\lambda+\tfrac{2}{3}\mu$, equation (4.99)). For a perfect monatomic gas $b=0$ but for other fluids b is generally not zero. It is important to realise from equation (13.8) that the effect of the bulk viscosity depends on the time rate of change of the density of the fluid and obviously it is very difficult to measure experimentally. It is only the ultrasonic absorption measurement which allows† one to infer the magnitude of b. For further discussion on the values of b, readers may be referred to Chapter 4 of Bhatia (1967).

The shear viscosity η (as in equation (13.1)), on the other hand, has been measured extensively for gases and liquids. Its value depends both on temperature and pressure. The coefficient η of a gas increases with temperature but η of a liquid, in general, decreases with the rise of temperature. Sometimes instead of shear viscosity η, one uses $\nu \equiv \eta/\rho$ which is called the kinematic viscosity. Table 13.1 gives some typical values of η for fluids.

Table 13.1 The shear viscosity† of typical fluids at atmospheric pressure.

Media	Temperature (°C)	η (P)
Air	18	1.82×10^{-4}
	1034	4.9×10^{-4}
Hydrogen	20.7	0.87×10^{-4}
	825	2.13×10^{-4}
Water	20	1.0×10^{-2}
	99	0.28×10^{-2}
Mercury	20.2	1.55×10^{-2}
	340	0.92×10^{-2}
Olive oil	20	0.84
	70	0.12
Glycerin	20	14.90
	30	6.29
Glucose	22	9.1×10^{13}
	100	2.5×10^{2}

† Data taken from Weast (1984).

13.3 The Navier–Stokes equation

We start with our master equation of motion (equation (6.8)) derived in Chapter 6, namely

$$\rho \frac{\mathrm{D}v_i}{\mathrm{D}t} = \rho F_i + \sum_j \frac{\partial \sigma_{ji}}{\partial x_j} \qquad i, j = 1, 2, 3$$

† See §14.3.

where various terms have their usual meaning. Substituting for σ from (13.2), the equations of motion for real fluids become

$$\rho \frac{\mathrm{D}v_i}{\mathrm{D}t} = \rho F_i + \sum_j \frac{\partial}{\partial x_j}(-p\delta_{ji} + \Pi_{ji}). \tag{13.10}$$

For $i=1$, one therefore has

$$\rho \frac{\mathrm{D}v_1}{\mathrm{D}t} = \rho F_1 - \frac{\partial p}{\partial x_1} + \frac{\partial \Pi_{11}}{\partial x_1} + \frac{\partial \Pi_{21}}{\partial x_2} + \frac{\partial \Pi_{31}}{\partial x_3}. \tag{13.11}$$

Since $\Pi_{11}, \Pi_{21}, \ldots$ are given through (13.4*b*), this reduces to

$$\rho \frac{\mathrm{D}v_1}{\mathrm{D}t} = \rho F_1 - \frac{\partial p}{\partial x_1} + \eta' \frac{\partial}{\partial x_1} \operatorname{div} \boldsymbol{v} + 2\eta \left(\frac{\partial v_{11}}{\partial x_1} + \frac{\partial v_{12}}{\partial x_2} + \frac{\partial v_{13}}{\partial x_3} \right). \tag{13.12}$$

The last bracketed terms can be further simplified in view of equation (13.3) and this leads to

$$\begin{aligned} \rho \frac{\mathrm{D}v_1}{\mathrm{D}t} &= \rho F_1 - \frac{\partial p}{\partial x_1} + (\eta + \eta') \frac{\partial}{\partial x_1} \operatorname{div} \boldsymbol{v} + \eta \nabla^2 v_1 \\ &= \rho F_1 - \frac{\partial p}{\partial x_1} + (b + \tfrac{1}{3}\eta) \frac{\partial}{\partial x_1} \operatorname{div} \boldsymbol{v} + \eta \nabla^2 v_1 \end{aligned} \tag{13.13}$$

or, in vector notation

$$\rho \frac{\mathrm{D}\boldsymbol{v}}{\mathrm{D}t} = \rho \boldsymbol{F} - \operatorname{grad} p + (b + \tfrac{1}{3}\eta) \operatorname{grad} \operatorname{div} \boldsymbol{v} + \eta \nabla^2 \boldsymbol{v}. \tag{13.14}$$

If the material derivative, $\mathrm{D}\boldsymbol{v}/\mathrm{D}t$, is expanded as in §10.2.2, i.e.

$$\frac{\mathrm{D}\boldsymbol{v}}{\mathrm{D}t} = \frac{\partial \boldsymbol{v}}{\partial t} + \tfrac{1}{2} \operatorname{grad} \boldsymbol{v}^2 - \boldsymbol{v} \times \operatorname{curl} \boldsymbol{v} \tag{13.15}$$

then equation (3.14) becomes

$$\begin{aligned} \rho \frac{\partial \boldsymbol{v}}{\partial t} &= \rho \boldsymbol{F} - \operatorname{grad} p - \tfrac{1}{2}\rho \operatorname{grad} \boldsymbol{v}^2 + \rho(\boldsymbol{v} \times \operatorname{curl} \boldsymbol{v}) \\ &\quad + (b + \tfrac{1}{3}\eta) \operatorname{grad} \operatorname{div} \boldsymbol{v} + \eta \nabla^2 \boldsymbol{v} \end{aligned} \tag{13.16}$$

Equations (13.16) or (13.14) are the celebrated Navier–Stokes (NS) equations of motion for real fluids. This equation forms the starting point for many investigations in the hydrodynamics and aerodynamics of real fluids. It is interesting to observe that the viscosity-containing terms in the NS equations have a similar structure to those in the equations of motion for isotropic solids†. We also notice that by deleting the viscosity-containing terms from

† For isotropic solids one has terms such as, $(\lambda + \mu) \operatorname{grad} \operatorname{div} \boldsymbol{s} + \mu \nabla^2 \boldsymbol{s}$ (see, for instance, equation (5.4)). If $\eta' \to \lambda$, $\eta \to \mu$ and $\boldsymbol{v} \to \boldsymbol{s}$ then the two become identical.

equation (13.16), one recovers Euler's equations of motion (10.15) for perfect fluids.

13.4 Energy dissipation due to viscosity

The viscosity of the fluid causes an irreversible loss of energy. The viscous forces are frictional forces which produce heat and are lost from the system. To quantify this let us recall expression (4.34) for the work done by the stress components σ_{ij} when the displacements of the particles at a point are changed quasi-statically (the system is in equilibrium at every instant of time) by $\Delta s_i(\boldsymbol{x})$ (i.e. $s_i(\boldsymbol{x}) \rightarrow s_i(\boldsymbol{x}) + \Delta s_i(\boldsymbol{x})$, $i = 1, 2, 3$). The amount of work done on the system per unit volume can be expressed as

$$\mathrm{d}w = \sum_{ij} \sigma_{ij} \left[\frac{1}{2} \left(\frac{\partial \, \Delta s_i}{\partial x_j} + \frac{\partial \, \Delta s_j}{\partial x_i} \right) \right].$$

Now suppose that the change in displacements Δs_i occur in a time δt so that

$$\Delta s_i = v_i \delta t \tag{13.17}$$

where $\boldsymbol{v}$ is the velocity of the fluid. We therefore obtain,

$$\frac{\delta w}{\delta t} = \sum_{ij} \sigma_{ij} v_{ij} \tag{13.18}$$

where v_{ij} is the velocity gradient tensor (see equation (13.3)). By substituting σ_{ij} from (13.5), this equation leads to

$$\frac{\delta w}{\delta t} = -p \operatorname{div} \boldsymbol{v} + \eta'(\operatorname{div} \boldsymbol{v})^2 + 2\eta \sum_{ij} v_{ij}^2. \tag{13.19}$$

We see that the rate at which work is being done per unit volume consists of two major terms, one is independent of viscosity and the other is viscosity dependent. The first represents the rate at which work is being done in changing the volume or density ($\operatorname{div} \boldsymbol{v} = -\rho^{-1} \mathrm{D}\rho/\mathrm{D}t$). It may be of either sign depending on $\operatorname{div} \boldsymbol{v}$. The viscosity-containing terms, on the other hand, are always positive (or at least never negative) and are responsible for the loss in energy. Denoting these by $(\delta\varepsilon/\delta t)_{\mathrm{vis}}$, one writes

$$\left(\frac{\delta \varepsilon}{\delta t} \right)_{\mathrm{vis}} = \eta'(\operatorname{div} \boldsymbol{v})^2 + 2\eta \sum_{ij} v_{ij}^2. \tag{13.20}$$

Simplification of equation (13.20) occurs in special cases:

(i) For incompressible fluids, $\operatorname{div} \boldsymbol{v} = 0$, therefore

$$\left(\frac{\delta \varepsilon}{\delta t} \right)_{\mathrm{vis}} = 2\eta \sum_{ij} v_{ij}^2. \tag{13.21}$$

(ii) If all v_{ij} are zero except v_{11} $(=\partial v_1/\partial x_1)$ then

$$\left(\frac{\delta \varepsilon}{\delta t}\right)_{\text{vis}} = (b+\tfrac{4}{3}\eta)\left(\frac{\partial v_1}{\partial x_1}\right)^2. \tag{13.22}$$

(iii) If all $v_{ij}=0$, except $v_{12}=v_{21}=\frac{1}{2}\partial v_1/\partial x_1$, then

$$\left(\frac{\delta \varepsilon}{\delta t}\right)_{\text{vis}} = \eta\left(\frac{\partial v_2}{\partial x_1}\right)^2. \tag{13.23}$$

These expressions will be useful later.

Problems

13.1 Write down the components of the stress tensor σ_{ij} for viscous fluids in cylindrical and spherical polar coordinates and hence derive the Navier–Stokes equation.

13.2 Using the full Navier–Stokes equation, show that for an incompressible fluid under conservative forces:

(i)
$$\frac{\partial \boldsymbol{\omega}}{\partial t} = \text{curl}\,(\boldsymbol{v}\times\boldsymbol{\omega}) + \frac{\eta}{\rho}\nabla^2\boldsymbol{\omega} \tag{P13.1}$$

(ii)
$$\frac{\mathrm{D}\Gamma}{\mathrm{D}t} = \frac{\eta}{\rho}\nabla^2\Gamma. \tag{P13.2}$$

$\boldsymbol{\omega}$ $(=\frac{1}{2}\,\text{curl}\,\boldsymbol{v})$ is the vorticity and Γ is the circulation.

14

SOUND PROPAGATION

In this Chapter we specialise the application of the Navier–Stokes equation to problems like sound propagation in real fluids. The propagation of sound waves, like other waves, causes a small perturbation in the medium. The density, the pressure and the temperature of an element of fluid vary with time. If the variation is completely reversible then there is no attenuation (energy loss) as one assumes for a perfect fluid. In a real fluid, however, the propagation is always associated with attenuation or absorption of energy. This might occur due to various causes[†]. The presence of viscosity and thermal conductivity of the medium are major sources for attenuation. This is usually known as classical absorption and presently we restrict our discussion to it.

14.1 Linearised Navier–Stokes equation

For easy reference we first recollect useful relations from previous Chapters:

(i) The material derivative, equation (6.5):

$$\frac{\mathrm{D}}{\mathrm{D}t}=\frac{\partial}{\partial t}+v_1\frac{\partial}{\partial x_1}+v_2\frac{\partial}{\partial x_2}+v_3\frac{\partial}{\partial x_3}.$$

(ii) The equation of continuity, equation (10.8*b*):

$$\mathrm{D}\rho/\mathrm{D}t+\rho \operatorname{div} \boldsymbol{v}=0.$$

(iii) The Navier–Stokes equation, equation (13.14):

$$\rho\mathrm{D}\boldsymbol{v}/\mathrm{D}t=\rho\boldsymbol{F}-\operatorname{grad} p+(b+\tfrac{1}{3}\eta)\operatorname{grad}\operatorname{div}\boldsymbol{v}+\eta\,\nabla^2\boldsymbol{v}.$$

We shall now see that the simplifications of the above relations are possible for problems like the propagation of sound waves. Let ρ_0 be the density of an element of fluid in absence of sound waves. When sound waves propagate, the medium experiences a perturbation and the density becomes $\rho=\rho_0+\delta\rho$, where $\delta\rho$ is the change in density due to $\boldsymbol{v}$. Since these waves are of weak intensity, $\delta\rho$ and $\boldsymbol{v}$ may be regarded as sufficiently small. The terms such as $\mathrm{O}(v^2)$, $\mathrm{O}(\delta\rho^2)$ and $\mathrm{O}(v\,\partial\rho/\partial x)$ may therefore be neglected. This is known as the

[†] For an extensive study see, e.g., Bhatia (1967) which may be recommended.

acoustic approximation. Thus, we express

$$\frac{\mathrm{D}\boldsymbol{v}}{\mathrm{D}t}=\frac{\partial \boldsymbol{v}}{\partial t}+v_j\frac{\partial \boldsymbol{v}}{\partial x_j}\simeq\frac{\partial \boldsymbol{v}}{\partial t}$$

$$\frac{\mathrm{D}\rho}{\mathrm{D}t}=\frac{\partial \rho}{\partial t}+v_j\frac{\partial \rho}{\partial x_j}\simeq\frac{\partial \rho}{\partial t}$$

$$\rho \operatorname{div} \boldsymbol{v}=\rho_0 \operatorname{div} \boldsymbol{v}.$$

The equation of continuity in the acoustical approximation, therefore, becomes

$$\frac{\partial \rho}{\partial t}+\rho_0 \operatorname{div} \boldsymbol{v}=0. \tag{14.1}$$

Similarly, if we write

$$\rho\frac{\mathrm{D}\boldsymbol{v}}{\mathrm{D}t}\simeq\rho\frac{\partial \boldsymbol{v}}{\partial t}\simeq\rho_0\frac{\partial \boldsymbol{v}}{\partial t}$$

$$\rho\boldsymbol{F}\simeq\rho_0\boldsymbol{F}$$

then the Navier–Stokes equation becomes

$$\rho_0\frac{\partial \boldsymbol{v}}{\partial t}=\rho_0\boldsymbol{F}-\operatorname{grad} p+(b+\tfrac{1}{3}\eta)\operatorname{grad}\operatorname{div}\boldsymbol{v}+\eta\,\nabla^2\boldsymbol{v}. \tag{14.2}$$

For problems like sound propagation the body force is zero ($\boldsymbol{F}=0$) equation (14.2) reduces to

$$\rho_0\frac{\partial \boldsymbol{v}}{\partial t}=-\operatorname{grad} p+(b+\tfrac{1}{3}\eta)\operatorname{grad}\operatorname{div}\boldsymbol{v}+\eta\,\nabla^2\boldsymbol{v}. \tag{14.3}$$

14.2 Sound propagation in the absence of viscosity

If all the viscosity terms are zero then equation (14.3) reduces to

$$\rho_0\,\partial\boldsymbol{v}/\partial t=-\operatorname{grad} p. \tag{14.4}$$

During the sound propagation we assume that the local entropy of an element remains constant and the pressure is a unique function of the density ($\rho=\rho_0+\delta\rho$). By expanding p in a Taylor's series and retaining only the linear terms, one has

$$p(\rho)=p(\rho_0+\delta\rho)=p(\rho_0)+\left(\frac{\delta p}{\delta\rho}\right)_{\rho=\rho_0}(\rho-\rho_0) \tag{14.5a}$$

or

$$\operatorname{grad} p=(\delta p/\delta\rho)\operatorname{grad}\rho \tag{14.5b}$$

where $(\delta p/\delta\rho)\equiv(\delta p/\delta\rho)_{\rho=\rho_0}$. In this way equation (14.4) becomes

$$\rho_0(\partial \mathbf{v}/\partial t)=-(\delta p/\delta\rho)\text{ grad }\rho \tag{14.6a}$$

or

$$\rho_0\left(\frac{\partial^2\mathbf{v}}{\partial t^2}\right)=-\left(\frac{\delta p}{\delta\rho}\right)\text{grad }\frac{\partial\rho}{\partial t}. \tag{14.6b}$$

By making a substitution for $\partial\rho/\partial t$ from equation (14.1) one obtains

$$\rho_0\left(\frac{\partial^2\mathbf{v}}{\partial t^2}\right)=\rho_0\left(\frac{\delta p}{\delta\rho}\right)\text{grad div }\mathbf{v}. \tag{14.7}$$

If we take the divergence of (14.7),

$$\frac{\partial^2 G}{\partial t^2}=\frac{\delta p}{\delta\rho}\nabla^2 G \tag{14.8}$$

where $G\equiv\text{div }\mathbf{v}$. Equation (14.8) represents a wave equation for the propagation of dilatational (irrotational) waves. The velocity of propagation is

$$C_0=\left(\frac{\delta p}{\delta\rho}\right)^{1/2}=\left(\frac{\text{Bulk modulus}}{\rho_0}\right)^{1/2}. \tag{14.9}$$

It can easily be seen that in the present case there is no propagation of rotational waves. By taking the curl of equation (14.6*a*), one readily obtains ($\boldsymbol{\omega}=\frac{1}{2}\text{ curl }\mathbf{v}$)

$$\rho_0\,\partial\boldsymbol{\omega}/\partial t=0$$

or

$$\boldsymbol{\omega}=\text{constant}$$

which shows that the time-dependent solutions for rotations need not be considered.

Let us assume a plane-wave solution of equation (14.8) travelling along the x_1-direction,

$$v_1=v_1^0\exp\left[i(\omega t-kx_1)\right]. \tag{14.10}$$

($v_2=v_3=0$ because curl $\mathbf{v}=0$.) ω is the angular frequency and $k=2\pi/\lambda$ is the wavenumber. Substitution of (14.10) into (14.8) implies

$$C_0=\frac{\omega}{k}=\left(\frac{\delta p}{\delta\rho}\right)^{1/2}. \tag{14.11}$$

At this point if we compute the intensity of sound waves, $\mathscr{I}$, we may examine the validity of the approximations made earlier for the linearisation of the Navier–Stokes equation. $\mathscr{I}$ is defined as the amount of energy passing per unit volume across a unit area normal to the direction of propagation and is given by the product of the sound energy in a unit volume and the velocity C_0. For plane waves (14.10), it may be expressed as

$$\mathscr{I} = [\tfrac{1}{2}\rho_0 (v_1^0)^2] C_0$$
$$= \tfrac{1}{2}\rho_0 C_0^3 (v_1^0/C_0)^2. \qquad (14.12)$$

Equation (14.12) can readily be used to compute (v_1^0/C_0) for a given intensity $\mathscr{I}$. The intensity of plane sound waves produced in the laboratory generally varies between 0.1 to 0.3 W cm^{-2}. The values of (v_1^0/C_0) corresponding to $\mathscr{I} =$ 0.1 W cm^{-2} for typical media are shown in table 14.1. In each case we see that $v_1^0 \ll C_0$ and therefore the particle velocity may be regarded as small as assumed earlier. It may, however, be noticed that this is valid only for sound waves of weak intensity. Obviously, for larger intensities non-linear terms in the equation of motion may not be neglected.

Table 14.1 Values of (v_1^0/C_0) in different media for sound waves of intensity 0.1 W cm^{-2}.

Media	ρ_0 (g cm^{-3})	C_0 (cm s^{-1})	v_1^0/C_0
Air	1.2×10^{-3}	3.4×10^4	7×10^{-3}
Water	1.0	1.5×10^5	2.4×10^{-5}
Copper	9.0	5.0×10^5	1.3×10^{-6}

14.3 Effect of viscosity on sound propagation

When a sound wave passes through a viscous medium then due to internal friction a part of the mechanical energy is converted into heat. This causes a loss or attenuation (also called absorption) of the wave energy. This is better explained with the help of the Navier–Stokes equation (14.3). The latter in association with equation (14.5*b*) can be expressed as

$$\rho_0 \frac{\partial \mathbf{v}}{\partial t} = -\left(\frac{\delta p}{\delta \rho}\right) \text{grad } \rho + (b + \tfrac{1}{3}\eta) \text{ grad div } \mathbf{v} + \eta \nabla^2 \mathbf{v}. \qquad (14.13)$$

By taking $\partial/\partial t$ on both sides and remembering the equation of continuity (14.1) one has

$$\rho_0 \frac{\partial^2 \mathbf{v}}{\partial t^2} = \rho_0 C_0^2 \text{ grad div } \mathbf{v} + \frac{\partial}{\partial t}[(b + \tfrac{1}{3}\eta) \text{grad div } \mathbf{v} + \eta \nabla^2 \mathbf{v}] \qquad (14.14)$$

where $C_0^2 = \delta p/\delta\rho, \rho_0$ is the density of the unperturbed medium, $\mathbf{v}$ is the particle velocity, b is the bulk viscosity and η is the shear viscosity. It is obvious from

(14.14) that unlike perfect fluids not only the longitudinal but shear waves[†] may also propagate through a viscous medium. Presently we are interested in the propagation of the longitudinal wave and therefore in one dimension (the particle velocity v_1 is assumed along x_1; $v_2=v_3=0$), equation (14.14) reduces to

$$\rho_0 \frac{\partial^2 v_1}{\partial t^2}=\rho_0 C_0^2 \frac{\partial^2 v_1}{\partial x_1^2}+(b+\tfrac{4}{3}\eta)\frac{\partial}{\partial t}\frac{\partial^2 v_1}{\partial x_1^2}. \qquad (14.15)$$

For convenience let us introduce an abbreviation

$$\omega_v=\frac{\rho_0 C_0^2}{b+\frac{4}{3}\eta}. \qquad (14.16)$$

ω_v is often known as the viscosity relaxation frequency. Equation (14.15) therefore becomes

$$\frac{\partial^2 v_1}{\partial t^2}=C_0^2\frac{\partial^2 v_1}{\partial x_1^2}+\frac{C_0^2}{\omega_v}\frac{\partial}{\partial t}\frac{\partial^2 v_1}{\partial x_1^2}. \qquad (14.17)$$

Consider a plane-wave solution

$$v_1=v_1^0 \exp\,[\mathrm{i}(\omega t-kx_1)] \qquad v_2=v_3=0.$$

By substituting this into (14.17) one obtains

$$\omega^2=C_0^2k^2(1+\mathrm{i}\omega/\omega_v) \qquad (14.18)$$

or

$$k^2=\frac{\omega^2}{C_0^2}\frac{1}{1+\mathrm{i}\omega/\omega_v} \qquad (14.19)$$

which is the desired relation between the frequency ω and the wavenumber k. The ω–k relation is also known as the dispersion relation. Since equation (14.19) is a complex relation, unlike the perfect fluid equation (14.11), it has a different physical interpretation. In order to interpret this let us take the wavenumber k to be a complex number, i.e.

$$k=k_1-\mathrm{i}k_2$$

where k_1 and k_2 are real and positive. These have the same physical significance as discussed earlier (see equation (8.60)): the real part k_1 ($=\omega/C$) provides the velocity of propagation and the imaginary part k_2 gives the attenuation. In this view, the plane-wave solution becomes

$$v_1=v_1^0 \exp\,(-k_2x_1)\exp\,[\mathrm{i}(\omega t-k_1x_1)]. \qquad (14.20)$$

We observe that the amplitude of the wave diminishes exponentially as the wave progresses along the x_1-direction. At a distance $x_1=1/k_2$ from the source

[†] We shall consider shear waves in the next section.

the amplitude reduces to 1/e (i.e. of the order of 0.367) of its original values. k_2 is called the amplitude attenuation per centimetre if the distance x_1 is measured in centimetres.

By writing $k=k_1-\mathrm{i}k_2$ in equation (14.19) one has

$$k_1^2-k_2^2-2\mathrm{i}k_1k_2=\frac{\omega^2}{C_0^2}\frac{1}{1+\mathrm{i}\omega/\omega_v}. \tag{14.21}$$

For low frequencies ($\omega \ll \omega_v$, which is usually the case), one may re-express (14.21) as

$$k_1^2-2\mathrm{i}k_1k_2 \simeq (\omega^2/C_0^2)(1-\mathrm{i}\omega/\omega_v). \tag{14.22}$$

Equation (14.22) follows because $\omega/\omega_v \ll 1$ and $k_2/k_1 \ll 1$. The separation of the real and imaginary parts yields

$$C^2=\omega^2/k_1^2=C_0^2 \tag{14.23}$$

and

$$k_2=\omega^2/2C_0\omega_v. \tag{14.24}$$

By substituting the value of ω_v from equation (14.16), it becomes

$$k_2=\frac{\omega^2}{2\rho_0C_0^3}(b+\tfrac{4}{3}\eta). \tag{14.25a}$$

Occasionally k_2, i.e. the amplitude attenuation per cm is also expressed in terms of the amplitude attenuation per wavelength, α,

$$\alpha=k_2\lambda=2\pi k_2/k_1.$$

Therefore (14.25*a*) may also be expressed as

$$\alpha=\frac{\pi\omega}{\rho_0C_0^2}(b+\tfrac{4}{3}\eta). \tag{14.25b}$$

Equations (14.23) and (14.25) are the required relations for the propagation velocity and for the attenuation of the wave energy due to viscosity in the low-frequency limit ($\omega \sim 10^7$ Hz). It may be observed from (14.23) that there is no velocity dispersion, i.e. the velocity of the propagation C remains unmodified, however, there is a loss in energy through equation (14.25). The latter may also be derived independently which is shown in Appendix (A.2). Some interesting observations may be made as follows.

(i) Bulk viscosity

Equation (14.25*a*) is very useful to determine† the bulk viscosity b using the experimentally observed attenuation. Due to this equation the attenuation k_2 seems to consist of two terms; one is due to bulk viscosity b and the other is due to shear viscosity η. In the absence of b, equation (14.25*a*) simply reduces to the

† There are no direct methods for the measurements of bulk viscosity.

Stokes attenuation formula,

$$k_2^{\text{Stokes}} = 2\omega^2\eta/3\rho_0 C_0^3. \tag{14.26}$$

From equation (14.25*a*) and (14.26) one has

$$\frac{b}{\eta} = \frac{4}{3}\left(\frac{k_2 - k_2^{\text{Stokes}}}{k_2^{\text{Stokes}}}\right). \tag{14.27}$$

Typical values for b/η for some liquids are given in table 14.2.

Table 14.2 Ratio $(b/\eta)^{\dagger}$ of bulk-to-shear viscosity at 20 °C.

Glycerol	Water	Methyl alcohol	Toluene	Benzene
1.1	2.5	3.2	13	100

† Data from Litovitz (1960).

(ii) Attenuation is proportional to the square of the frequency
From equation (14.25*a*), the attenuation of the wave energy is proportional to the square of the sound frequency. For a large wave energy there is more attenuation. This immediately reminds us that if sound waves are being used as a probe for investigating the physical properties of matter then one should use as weak an intensity as possible. The square-law relation between the attenuation and the frequency is found to be experimentally true for the majority of liquids and gases so long as the frequencies are less or of the order of 10^7 Hz.

(iii) Highly viscous liquids
For highly viscous liquids, i.e. for larger values of η, ω_v becomes smaller or of the order of ω. The condition $\omega/\omega_v \ll 1$ is not satisfied and hence equation (14.25*a*) can no longer be used here. In such cases equation (14.21) can be solved exactly to obtain $C\ (=\omega/k_1)$ and k_2,

$$C^2 = 2C_0^2 \frac{1+(\omega/\omega_v)^2}{1+[1+(\omega/\omega_v)^2]^{1/2}} \tag{14.28}$$

$$k_2 = \frac{\omega C}{2C_0^2}\frac{\omega/\omega_v}{1+(\omega/\omega_v)^2}. \tag{14.29}$$

These functions are plotted in figures 14.1 and 14.2 respectively and are compared with the experimental observations for glycerol. The disagreement between the theoretical predictions and the experimental observations is quite obvious and thus underlines the fact that the Navier–Stokes equation is not valid for such systems. As a matter of fact this was expected because it may be seen from the structure of the NS equation (14.15) that the latter is not useful

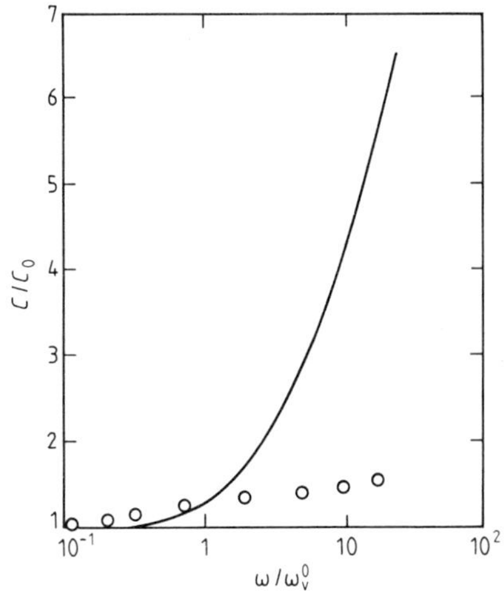

Figure 14.1 Sound velocity in glycerol (5% water). The full curve is the result calculated from the Navier–Stokes equation (14.28) (ω_v^0 in the horizontal scale refers to ω_v for $b=0$) and the circles denote experimental points. (Source, Litovitz and Sette (1953), see also Bhatia (1967).)

for highly viscous liquids. The right-hand side of equation (14.15) consists of two terms; the first is independent of viscosity but the second term depends on viscosity and contains higher derivatives of $\boldsymbol{v}$ than the former. Earlier in writing the viscosity stress tensor all higher-order derivatives in $\boldsymbol{v}$ are neglected. By the same token the second term in equation (14.15) may be expected to be smaller than the first, i.e.

$$\left|\frac{C_0^2}{\omega_v}\frac{\partial}{\partial t}\frac{\partial^2 v_1}{\partial x_1^2}\right| \ll \left|C_0^2\frac{\partial^2 v_1}{\partial x_1^2}\right|. \tag{14.30}$$

For a plane harmonic wave the inequality reduces to

$$\omega/\omega_v \ll 1 \tag{14.31}$$

which is a built-in limitation for the Navier–Stokes equation.

14.4 Diffusion of vorticity and viscous waves

At this stage we make a small digression from longitudinal waves, discussed above, and consider the rotational motion or vorticity in a fluid. The latter gives rise to shear waves which are usually known as viscous waves. We first recall the linearised Navier–Stokes equation (14.14),

$$\rho_0\frac{\partial^2\boldsymbol{v}}{\partial t^2}=\rho_0 C_0^2\,\text{grad div}\,\boldsymbol{v}+\frac{\partial}{\partial t}\,[(b+\tfrac{1}{3}\eta)\,\text{grad div}\,\boldsymbol{v}+\eta\,\nabla^2\boldsymbol{v}].$$

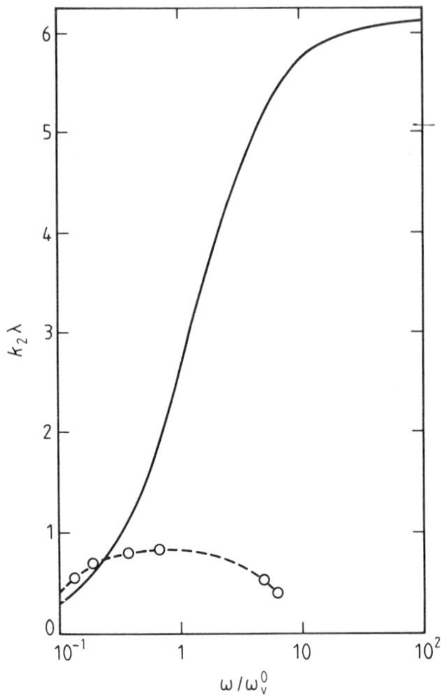

Figure 14.2 Attenuation per wavelength in glycerol (5% water). The full curve is the result calculated from the Navier–Stokes equation (14.29) (ω_v^0 has the same meaning as in figure 14.1) and the circles represent the experimental points. (Source, Litovitz and Sette (1953), see also Bhatia (1967).)

By taking the curl of both sides and remembering that curl grad = 0, one has

$$\partial \boldsymbol{\omega}/\partial t = (\eta \rho_0^{-1})\, \nabla^2 \boldsymbol{\omega} \tag{14.32}$$

where $\boldsymbol{\omega} = \frac{1}{2}$ curl $\boldsymbol{v}$ is the vorticity. Equation (14.32) may be called the diffusion equation for the vorticity $\boldsymbol{\omega}$.

Consider a transverse-wave solution such that the wave motion is directed along the x_1-axis and the particle motion is along the x_2-axis, i.e.

$$v_2 = v_2^0 \exp\,[\mathrm{i}(\omega t - kx_1)] \qquad v_1 = v_3 = 0. \tag{14.33}$$

Obviously div $\boldsymbol{v} = 0$ here and therefore equation (14.14) reduces to

$$\frac{\partial^2 v_2}{\partial t^2} = \frac{\eta}{\rho_0}\frac{\partial}{\partial t}\left(\frac{\partial^2 v_2}{\partial x_1^2}\right). \tag{14.34}$$

Substitution of (14.33) into (14.34) yields the dispersion relations,

$$k^2 = -\mathrm{i}\omega\rho_0/\eta. \tag{14.35}$$

By setting $k=k_1-ik_2$, one has

$$k_1^2-k_2^2-2ik_1k_2=-i\omega\rho_0/\eta. \tag{14.36}$$

The separation of real and imaginary parts provides,

$$k_1=k_2 \tag{14.37a}$$

and

$$k_1k_2=\omega\rho_0/2\eta. \tag{14.37b}$$

The attenuation k_2 and the propagation velocity C of the viscous waves, therefore, becomes,

$$k_2=(\omega\rho_0/2\eta)^{1/2} \tag{14.38a}$$

$$C=\frac{\omega}{k_1}=\left(\frac{2\eta\omega}{\rho_0}\right)^{1/2}. \tag{14.38b}$$

As a further illustration, we consider the propagation of such waves through water: $\eta=10^{-2}$ P, $\rho_0=1.0$ g cm^{-3} and $\nu=\omega/2\pi=10^5$ Hz;

$$C=(4\pi\times10^{-2}\times10^5)^{1/2}\simeq1.1\times10^2 \text{ cm s}^{-1} \tag{14.39}$$

$$k_2=\left(\frac{2\pi\times10^5}{2\times10^{-2}}\right)\simeq0.56\times10^4 \text{ cm}. \tag{14.40}$$

Thus we observe that the viscous waves are highly damped. For liquids with large η these waves are, however, less damped.

14.5 General equation of heat transfer and the effect of heat flow on wave propagation

Like viscosity, heat flow is another important source for the attenuation of sound waves in monatomic fluids. As is well known the propagation of sound waves is accompanied with the compression and rarefaction of the medium. The compressed region experiences a rise in temperature and thus heat flows from here to the rarefied regions. This contributes towards an increase in the entropy of the system and hence some irreversible loss in energy takes place. To investigate this, we need to couple the equations of motion with the thermodynamical variables.

14.5.1 Coupling of equations of motion with thermodynamics

From the combined statement of the first and second laws of thermodynamics (see equation (8.4)),

$$dQ=T\,dS=dU+dW$$

where dQ is the amount of heat supplied to the system, dS is the change in entropy, dU is the change in internal energy and dW represents the work done by the system. All quantities refer to unit volume. For fluids we can write $dW = p\,dV$, then

$$T\,dS = dU + p\,dV.$$

We may also express $T\,dS$ as

$$T\,dS = dQ - dW + p\,dV. \tag{14.41}$$

For viscous fluids, the work done on unit volume of the system in time dt has been calculated earlier (see equation (13.19)), i.e.

$$dW = (-p \operatorname{div} \boldsymbol{v})\,dt + \Pi_{ij} v_{ij}\,dt$$

where Π_{ij} $(=\eta' \operatorname{div} \boldsymbol{v}\delta_{ij} + 2\eta v_{ij})$ are components of the viscosity stress tensor. Therefore the work done by the system may be expressed as

$$dW = (p \operatorname{div} \boldsymbol{v})\,dt - \Pi_{ij} v_{ij}\,dt. \tag{14.42}$$

By substituting (14.42) into (14.41) and bearing in mind that the first term on the right-hand side of (14.42) is simply $p\,dV$, one has

$$T\,dS = dQ + \Pi_{ij} v_{ij}\,dt. \tag{14.43}$$

More appropriately if we express S in terms of the material derivative then (14.43) becomes

$$T\frac{DS}{Dt} = \frac{dQ}{dt} + \Pi_{ij} v_{ij}. \tag{14.44}$$

or

$$T\left(\frac{\partial S}{\partial t} + \boldsymbol{v}\cdot \operatorname{grad} S\right) = \frac{dQ}{dt} + \Pi_{ij} v_{ij}. \tag{14.45}$$

This represents a general equation for the heat transfer. Before we could apply (14.45) to a real problem, it is necessary to determine dQ. Heat may enter or leave an element of volume due either to radiation or heat conduction or due to both. Of these, heat radiation is usually of importance at very high temperatures (~ 1000 K) as in astrophysics problems. For the present purpose it is sufficient to consider the heat flow only due to heat conduction. The rate at which the heat is flowing out through a cross sectional area A is given by

$$dQ/dt = -\chi A \operatorname{grad} T \tag{14.46}$$

where χ is the thermal conductivity. The heat flowing in per unit volume therefore becomes

$$dQ/dt = \operatorname{div}(\chi \operatorname{grad} T). \tag{14.47}$$

By taking equations (14.45) and (14.47) together, we obtain

$$T(\partial S/\partial t+\boldsymbol{v}\cdot\text{grad } S)=\text{div }(\chi \text{ grad } T)+\Pi_{ij}v_{ij} \tag{14.48}$$

where S refers to unit volume of the fluid. If we refer it to unit mass of the fluid then (14.48) is

$$\rho T(\partial S/\partial t+\boldsymbol{v}\cdot\text{grad } S)=\text{div }(\chi \text{ grad } T)+\Pi_{ij}v_{ij}. \tag{14.49}$$

Equation (14.49) is a useful form of the heat transfer equation. It simplifies considerably, however, in special cases.

For an incompressible fluid the thermodynamic pressure remains the same and therefore we may write,

$$T\frac{\partial S}{\partial t}=T\left(\frac{\partial S}{\partial T}\right)_P\frac{\partial T}{\partial t}=C_P\frac{\partial T}{\partial t} \tag{14.50}$$

$$\text{grad } S=\left(\frac{\partial S}{\partial t}\right)_P\text{grad } T=\frac{C_P}{T}\text{grad } T \tag{14.51}$$

where C_P is the specific heat at constant pressure. In view of these, equation (14.49) becomes

$$\left(\frac{\partial T}{\partial t}+\boldsymbol{v}\cdot\text{grad } T\right)=\frac{1}{\rho C_P}[\text{div }(\chi \text{ grad } T)+\Pi_{ij}v_{ij}] \tag{14.52}$$

which is the appropriate equation to study the temperature distribution in a moving viscous fluid. For fluids at rest it reduces to the usual diffusion equation, i.e.

$$\frac{\partial T}{\partial t}=\frac{1}{\rho C_P}\text{div }(\chi \text{ grad } T). \tag{14.53}$$

Since χ is independent of the space variables, we can write

$$\frac{\partial T}{\partial t}=\frac{\chi}{\rho C_P}\nabla^2 T. \tag{14.54}$$

Another simplified version of (14.49) appears for the adiabatic motion of perfect fluids. Neglecting the effect of viscosity and thermal conductivity, it reads

$$\frac{\partial S}{\partial t}+\boldsymbol{v}\cdot\text{grad } S=0 \tag{14.55}$$

or

$$\partial(\rho S)/\partial t+\text{div }(\rho S\boldsymbol{v})=0. \tag{14.56}$$

Equation (14.56) is often known as the continuity equation for the entropy of a perfect fluid.

14.5.2 Effect of thermal conductivity on sound waves

We shall now use the heat transfer equation (14.49) to study the effect of heat flow on the propagation of sound waves in fluids. For simplicity we consider the effect of heat flow alone, i.e. the viscosity effect will be ignored. In the linear approximation the terms ($\boldsymbol{v}\cdot\text{grad}\ S$) and ($\Pi_{ij}v_{ij}$) in equation (14.49) may be set equal to zero and so we have,

$$T\,\partial S/\partial t=\rho_0^{-1}\ \text{div}\ (\chi\ \text{grad}\ T)$$

$$=\chi\rho_0^{-1}\ \nabla^2 T. \tag{14.57}$$

By using the thermodynamic relation,

$$T\ \mathrm{d}S=C_V\ \mathrm{d}T-\frac{T}{\rho_0^2}\left(\frac{\partial P}{\partial T}\right)_\rho \mathrm{d}\rho \tag{14.58}$$

equation (14.57) becomes

$$\frac{\chi}{\rho_0}\ \nabla^2 T=C_V\frac{\partial T}{\partial t}-\frac{T}{\rho_0^2}\left(\frac{\partial P}{\partial T}\right)_\rho\left(\frac{\partial \rho}{\partial t}\right). \tag{14.59}$$

On the other hand, employing equations (14.4) and (14.1), the equation of motion for non-viscous fluids becomes

$$\partial^2\rho/\partial t^2=\nabla^2 p. \tag{14.60}$$

For $p\equiv P$, the thermodynamic pressure, we have

$$\partial^2\rho/\partial t^2=\nabla^2 P. \tag{14.61}$$

Since P is a function of ρ and T, we write

$$\mathrm{d}P=\left(\frac{\partial P}{\partial T}\right)_\rho \mathrm{d}T+\left(\frac{\partial P}{\partial \rho}\right)_T \mathrm{d}\rho. \tag{14.62}$$

In view of this, equation (14.61) may be re-expressed as,

$$\frac{\partial^2\rho}{\partial t^2}=\left(\frac{\partial P}{\partial T}\right)_\rho \nabla^2 T+\left(\frac{\partial P}{\partial \rho}\right)_T \nabla^2\rho. \tag{14.63}$$

Equations (14.63) and (14.59) are the required relations describing the heat flow (thermal conductivity) in a perfect fluid due to the propagation of sound waves. Let the associated density and temperature variations be represented as,

$$\rho=\rho_0+\delta\rho\ \exp\ [\mathrm{i}(\omega t-kx)] \tag{14.64}$$

$$T=T_0+\delta T\ \exp\ [\mathrm{i}(\omega t-kx)] \tag{14.65}$$

where ρ_0 and T_0 are the density and temperature in the absence of sound waves. By substituting these in equations (14.59) and (14.63) we obtain

$$\left(\frac{\chi k^2}{\rho_0}+\mathrm{i}\omega C_V\right)\delta T-\mathrm{i}\omega\frac{T}{\rho_0^2}\left(\frac{\partial P}{\partial T}\right)_\rho\delta\rho=0 \tag{14.66}$$

$$k^2\left(\frac{\partial P}{\partial T}\right)_\rho\delta T-\left[\omega^2-k^2\left(\frac{\partial P}{\partial \rho}\right)_T\right]\delta\rho=0. \tag{14.67}$$

By eliminating δT and $\delta\rho$, and making use of the thermodynamic relation,

$$C_P-C_V=\frac{T}{\rho_0^2}\left(\frac{\partial P}{\partial T}\right)_\rho^2\left(\frac{\partial P}{\partial \rho}\right)_T^{-1} \tag{14.68}$$

we obtain,

$$k^2=\frac{\omega^2}{(\partial P/\partial\rho)_T}\left(\frac{C_V-\mathrm{i}k^2\Psi}{C_P-\mathrm{i}k^2\Psi}\right) \tag{14.69}$$

where

$$\Psi=\chi/\rho_0\omega. \tag{14.70}$$

Equation (14.69) is the required dispersion relation. To obtain the velocity and attenuation of sound waves, we assume, as earlier, that

$$k^2=(k_1-\mathrm{i}k_2)^2\simeq k_1^2-2\mathrm{i}k_1k_2 \qquad \text{for } k_2^2\ll k_1^2. \tag{14.71}$$

By making the approximation that

$$\mathrm{i}k^2\Psi\simeq\mathrm{i}k_1^2\Psi=\mathrm{i}\left(\frac{\omega^2}{C^2}\right)\frac{\chi}{\rho_0\omega}=\frac{\mathrm{i}C_P\omega}{\omega_\chi} \tag{14.72}$$

where

$$\omega_\chi=C^2\rho_0C_P/\chi \tag{14.73}$$

equation (14.69) now becomes

$$k_1^2-2\mathrm{i}k_1k_2=\frac{\omega^2}{\gamma(\partial P/\partial\rho)_T}\left(\frac{1-\mathrm{i}\gamma\omega/\omega_\chi}{1-\mathrm{i}\omega/\omega_\chi}\right) \qquad \gamma=\frac{C_P}{C_V}. \tag{14.74}$$

The separation of real and imaginary parts provides

$$C^2=\frac{\omega}{k_1^2}=\gamma\left(\frac{\partial P}{\partial\rho}\right)_T\frac{1+\omega^2/\omega_\chi^2}{1+\gamma\omega^2/\omega_\chi^2} \tag{14.75}$$

$$k_2=\frac{C(\gamma-1)}{2\gamma(\partial P/\partial\rho)_T}\frac{\omega^2/\omega_\chi}{(1+\omega^2/\omega_\chi^2)}. \tag{14.76}$$

Equations (14.75) and (14.76) represent, respectively, the velocity dispersion and the attenuation of sound waves due to the thermal conduction in non-viscous fluids. Since the discussion of the above equations is based on the value ω/ω_χ, it is instructive first to find the value of ω_χ. The order of ω_χ computed

Table 14.3 Values of ω_χ (rad s^{-1}) for typical substances at room temperature.

Air	Water	Copper
$\sim 5 \times 10^9$	$\sim 1.6 \times 10^{13}$	$\sim 1.2 \times 10^{11}$

from the measured C, ρ_0, C_P and χ through equation (14.73) for typical substances are tabulated in table 14.3.

From expression (14.75) it is obvious that the sound propagation takes place adiabatically at low frequencies ($\omega \ll \omega_\chi$):

$$C^2 \simeq \gamma \left(\frac{\partial P}{\partial \rho}\right)_T = \frac{\gamma}{\rho_0}\left(-V\frac{\partial P}{\partial V}\right)_T = \gamma B_T \rho_0^{-1}.$$

The propagation of sound at high frequencies ($\omega \gg \omega_\chi$), on the other hand, is an isothermal process:

$$C^2 \simeq \left(\frac{\partial P}{\partial \rho}\right)_T = B_T \rho_0^{-1}.$$

As equation (14.76) indicates, the attenuation k_2 for $\omega \ll \omega_\chi$ is given by:

$$k_2 \simeq \frac{\chi}{2CC_0^2\rho_0 C_P}(\gamma - 1)\omega^2 \qquad (14.77)$$

which is proportional to the square of the frequency, which is a similar dependence to that of the viscosity. Also, $k_2 \propto \gamma - 1$, which shows that the attenuation due to thermal conductivity in solids is smaller than in fluids because $\gamma_{\text{solid}} < \gamma_{\text{fluid}}$.

The attenuation of sound due to thermal conductivity (i.e. equation (14.77)), and due to viscosity (see equation (14.25*a*)) are additive, i.e. $k_2^{\text{th}} + k_2^{\text{vis}} = k_2^{\text{class}}$, which is called the classical absorption. For monatomic fluids the classical values are found to be in good agreement with the experimental observation. However, for polyatomic fluids the observed attenuation is substantially greater than the classical values. This happens because for polyatomic fluids the characteristics of the molecule come into play and contribute significantly to the process of attentuation. For instance, the sound waves disturb the equilibrium distribution of energy between the internal (i.e. rotational and vibrational) and the external (i.e. translational) degrees of freedom of a molecule. Such a phenomenon is called thermal relaxation and forms a major source of attenuation in polyatomic fluids. The spatial arrangement of molecules at a given place may also be affected by the propagation of sound waves; the loss in energy due to this is called the structural relaxation. These aspects of attenuation will not be considered here, however, reference books such as Bhatia (1967) may be found useful for further work.

Problems

14.1 Verify expressions (14.28) and (14.29).

14.2 Neglect the term curl ($\boldsymbol{v} \times \boldsymbol{w}$) in equation (P13.1) and consider the plane-wave solution,

$$\boldsymbol{w} = \boldsymbol{w}_0 \exp [i(\omega t - kx)] \qquad \text{(P14.1)}$$

with $\boldsymbol{w}_0$ as a constant vector. ω is the (real) angular frequency and k is the complex wavenumber.

(i) Show that the velocity of the wave is

$$C = (2\eta\omega/\rho)^{1/2}. \qquad \text{(P14.2)}$$

(ii) Show that the amplitude of the wave attenuates to 1/e of its initial value in a distance $(2\eta/\omega\rho)^{1/2}$.

14.3 According to a formula due to Eucken, for an ideal gas,

$$\chi/\eta C_V = \tfrac{1}{4}(9\gamma - 5) \qquad \gamma = C_P/C_V. \qquad \text{(P14.3)}$$

Determine the ratio of sound attenuation due to heat conductivity and shear viscosity, the sound frequency being such that $\omega \ll \omega_\chi$, or $\omega \ll \omega_v$. What is the numerical value of this ratio for gaseous argon and air?

15

THE NAVIER–STOKES EQUATION: FLOW PROBLEMS

Finally we wish to address briefly some simple flow problems of an incompressible but real fluid. The simplified versions of the Navier–Stokes (NS) equation have been used to study the laminar flow leading to the Hagen–Poiseuille equation. The flow between two parallel plates and flow past a sphere (Stokes formula) have also been introduced. We conclude with a discussion on the turbulent flow based on Reynolds number.

15.1 Boundary conditions and the simplified version of the NS equation

The flow problems of the perfect fluid (i.e. without viscosity) have been discussed earlier in Chapters 11 and 12 on the basis of Euler's equation of motion. The necessary boundary conditions are given in §11.1.2. Presently we have the Navier–Stokes equation describing the motion of viscous fluids. We recall that NS is a differential equation with a higher order than Euler's equation, so we require more boundary conditions to solve it for a real physical problem. Based on experimental information it has been propounded that when a real fluid passes a fixed solid object then at the surface of the solid not only is the normal component (v_n) of the fluid velocity zero but also the tangential component (v_t) must vanish. That is to say, there is no relative motion between the solid and the fluid in contact (the fluid layer in contact with a fixed solid surface is at rest). For moving solid surfaces both v_n and v_t of the fluid are equal to v_n and v_t of the solid respectively.

In order to avoid complexities, we confine our discussion to viscous and incompressible ($\text{div}\,\boldsymbol{v}=0$) fluids in the absence of external forces ($\boldsymbol{F}=0$). The Navier–Stokes equation (13.14) then simplifies to

$$\rho \frac{\mathrm{D}\boldsymbol{v}}{\mathrm{D}t} = -\text{grad}\; p + \eta\, \nabla^2 \boldsymbol{v} \tag{15.1}$$

where ρ is the density of the fluid, $\boldsymbol{v}$ is the fluid velocity, p is the pressure and η is the shear viscosity.

15.2 Laminar flow

The exact solution of even the simplified version of the NS equation (15.1) is very difficult. We obtain its solution only in special cases. A particularly simple but important example is the flow known as laminar or streamline flow: an ordered flow wherein fluid particles move in parallel straight lines. The steady flow (not very fast) of a liquid in a straight capillary tube, or between two parallel plates and other similar flows are examples of laminar flow.

15.2.1 Steady state flow through a capillary tube (Hagen–Poiseuille equation)

Consider a laminar flow through a tube of length l and radius a where the end pressures are p_1 and p_2. The flow velocity $\mathbf{v}$ is taken along the axis of the tube, say the z-axis† (see figure 15.1). Since the flow is laminar, we must have

$$v_z \equiv v_z(r) \qquad v_x = v_y = 0. \tag{15.2}$$

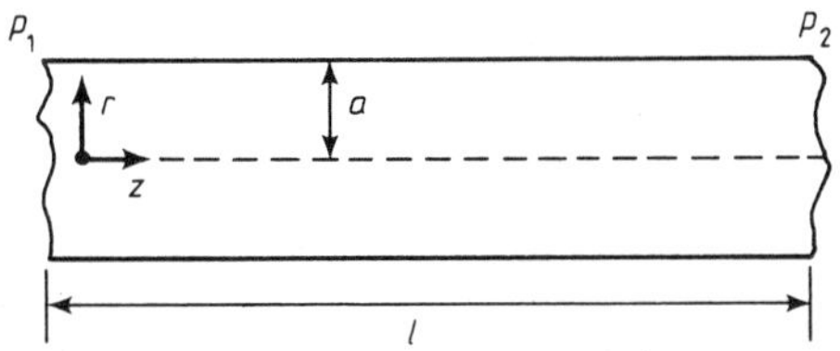

Figure 15.1 Laminar flow in a capillary tube.

For steady flow ($\partial v/\partial t = 0$), the condition (15.2) enables us to express the material derivative as,

$$\frac{\mathrm{D}\mathbf{v}}{\mathrm{D}t} = \frac{\partial \mathbf{v}}{\partial t} + v_x \frac{\partial \mathbf{v}}{\partial x} + v_y \frac{\partial \mathbf{v}}{\partial y} + v_z \frac{\partial \mathbf{v}}{\partial z} = 0. \tag{15.3}$$

Under this condition equation (15.1) reduces to

$$\eta\, \nabla^2 \mathbf{v} = \mathrm{grad}\ p. \tag{15.4}$$

If we make use of the cylindrical coordinates (r, φ, z), equation (15.4) becomes

$$\frac{\eta}{r}\left[\frac{\partial}{\partial r}\left(r\frac{\partial \mathbf{v}}{\partial r}\right)\right] + \frac{\eta}{r}\frac{\partial}{\partial \varphi}\left(\frac{1}{r}\frac{\partial \mathbf{v}}{\partial \varphi}\right) + \eta\frac{\partial}{\partial z}\left(\frac{\partial \mathbf{v}}{\partial z}\right) = \left(\frac{\partial}{\partial r} + \frac{1}{r}\frac{\partial}{\partial \varphi} + \frac{\partial}{\partial z}\right) p \tag{15.5}$$

and equivalent to (15.2) we have,

$$v_z = v_z(r) \qquad v_r = v_\varphi = 0. \tag{15.6}$$

† Note that the Cartesian axes x_1, x_2, x_3 are denoted here as the axes x, y, z respectively.

In view of this and remembering that p is independent of r and φ, equation (15.5) becomes

$$\frac{\eta}{r}\frac{\mathrm{d}}{\mathrm{d}r}\left(r\frac{\mathrm{d}v_z(r)}{\mathrm{d}r}\right)=\frac{\mathrm{d}p}{\mathrm{d}z}. \tag{15.7}$$

We observe that the left-hand side of this equation is a function of r only while the right-hand side is a function of z only, i.e. both sides must be a constant. This means that $\mathrm{d}p/\mathrm{d}z$ is a constant, and therefore we take,

$$\frac{\mathrm{d}p}{\mathrm{d}z}=\frac{p_2-p_1}{l}=-\frac{\Delta p}{l}. \tag{15.8}$$

Equation (15.7) now becomes

$$\frac{\mathrm{d}}{\mathrm{d}r}\left(r\frac{\mathrm{d}v_z(r)}{\mathrm{d}r}\right)=-\frac{\Delta p}{\eta l}r. \tag{15.9}$$

Integrating both sides we obtain

$$r\frac{\mathrm{d}v_z(r)}{\mathrm{d}r}=A-\frac{\Delta p}{2\eta l}r^2$$

or

$$\frac{\mathrm{d}v_z(r)}{\mathrm{d}r}=\frac{A}{r}-\frac{\Delta p}{2\eta l}r \tag{15.10}$$

where A is a constant of integration. If we integrate expression (15.10) once more, we obtain v_z, i.e.

$$v_z=B+A\ln r-(\Delta p/4\eta l)r^2 \tag{15.11}$$

where B is another constant of integration. A and B are determined from the boundary conditions. For $r=0$ (r is the distance from the axis of the tube), v_z from equation (15.11) becomes infinite due to the term $\ln r$. The velocity at the centre ($r=0$) of the tube however must remain finite; this is possible only if $A=0$. Also since $v_z=0$ at $r=a$ (radius of tube),

$$B=a^2\,\Delta p/4\eta l. \tag{15.12}$$

Therefore equation (15.11) becomes

$$v_z=\frac{\Delta p}{4\eta l}(a^2-r^2). \tag{15.13}$$

Equation (15.13) represents the velocity distribution in the tube which is parabolic in nature (see figure 15.2). The velocity is maximum along the axis and drops to zero at the inner surface of the tube. The maximum velocity along the axis ($r=0$) is

$$v_z^{\max}=a^2\,\Delta p/4\eta l. \tag{15.14}$$

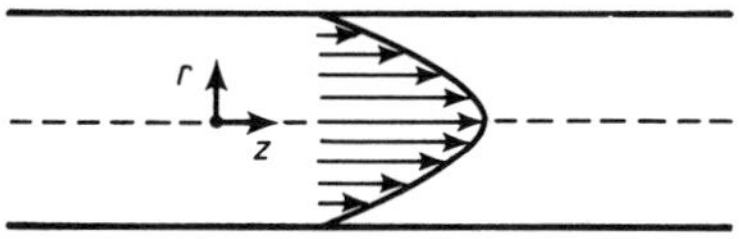

Figure 15.2 Velocity distribution in a laminar flow in a tube.

The mean velocity, v_0, of the fluid through any cross section of the tube may be obtained from

$$v_0 = \int_0^a 2\pi r v_z \, dr \left(\int_0^a 2\pi r \, dr \right)^{-1}. \tag{15.15}$$

With v_z as in (15.13), this can be integrated to give

$$v_0 = \frac{a^2}{8\eta} \frac{\Delta p}{l} \tag{15.16a}$$

$$= \tfrac{1}{2} v_z^{\max}. \tag{15.16b}$$

Thus for a given a and η, the mean velocity $v_0 \propto \Delta p$. This is, however, true for v_0 not exceeding a certain limit. Beyond this limit the flow no longer remains laminar but rather it becomes turbulent. Reynolds performed a series of careful experiments and set the limiting criteria for the validity of the laminar flow. We shall come to this again in the next section.

Equation (15.13) may readily be employed to calculate the mass Q of the fluid crossing each cross section of the tube per second. Q is often called the discharge and is given by

$$Q = \int_0^a \rho 2\pi r v_z \, dr. \tag{15.17}$$

By substituting v_z from (15.13) and integrating one obtains

$$Q = \pi a^4 \rho \, \Delta p / 8\eta l. \tag{15.18}$$

Thus for given values of $\Delta p/l$ and η, the discharge Q is proportional to the fourth power of the radius of the tube. Equation (15.18) is known as the Hagen–Poiseuille formula which is one of the best verified relations in hydrodynamics and is found to agree well with experiment within the limit of laminar flow. It is often used to determine the shear viscosity η.

15.2.2 Flow between two parallel plates

Consider two parallel plates at $x = 0$ and $x = h$ as shown in figure 15.3. The upper plate is moving with a velocity V parallel to the z-axis while the lower plate is at rest. Let a liquid be flowing between these two plates along the z-axis with velocity $v = v_z$ such that

$$v_x = v_y = 0 \qquad v_z = v_z(x). \tag{15.19}$$

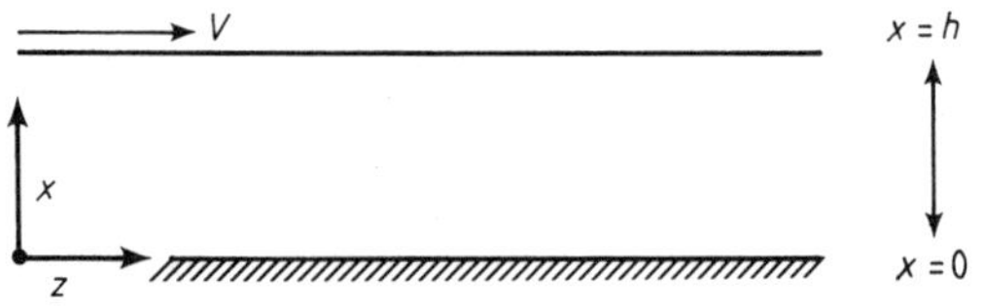

Figure 15.3 Laminar flow between two parallel plates. The upper plate is moving with velocity V and the lower plate is at rest.

Obviously the flow velocity v_z is constant in the direction of motion but it has a gradient in the x-direction.

For the steady-state flow $\mathrm{D}v/\mathrm{D}t=0$ and therefore the NS equation (15.1) reduces to

$$\eta \frac{\mathrm{d}^2 v_z}{\mathrm{d}x^2}=\frac{\mathrm{d}p}{\mathrm{d}z}. \tag{15.20}$$

This is similar to equation (15.7) of which the left-hand side is a function of x only and the right-hand side is a function of z only. By taking $\mathrm{d}p/\mathrm{d}z$ as a constant, equation (15.20) integrates out to give

$$v_z=A+Bx+\frac{x^2}{2\eta}\frac{\mathrm{d}p}{\mathrm{d}z} \tag{15.21}$$

where A and B are integration constants. The boundary condition at $x=0$, $v_z=0$, implies that $A=0$. On the other hand, at $x=h$, $v_z=V$, i.e. the velocity of the liquid in contact with the upper plate is simply the velocity of the plate, we have

$$B=\frac{1}{h}\left(V-\frac{h^2}{2\eta}\frac{\mathrm{d}p}{\mathrm{d}z}\right). \tag{15.22}$$

By making substitutions for A and B, equation (15.21) becomes

$$v_z=\frac{Vx}{h}+\frac{1}{2\eta}\frac{\mathrm{d}p}{\mathrm{d}z}(x^2-hx). \tag{15.23}$$

This is the required expression for the flow velocity between two parallel plates. Such a flow is sometimes known as the generalised Couette flow. It may be observed that even in the absence of the pressure gradient, $\mathrm{d}p/\mathrm{d}z=0$, the flow is possible just because of the motion of the upper plate. With $\mathrm{d}p/\mathrm{d}z=0$, equation (15.23), however, reduces to the simple Couette flow, i.e.

$$v_z=Vx/h \tag{15.24}$$

By employing equations (15.23) and (13.4), the tangential stress or the drag force per unit area on the two plates can immediately be calculated. From

(13.4) (consider only shear viscosity) one has

$$\begin{aligned}\Pi_{xz} &= 2\eta v_{xz} \\ &= \eta \frac{\partial v_z}{\partial x} \\ &= \frac{\eta V}{h} + \frac{1}{2}\frac{\mathrm{d}p}{\mathrm{d}z}(2x-h).\end{aligned} \tag{15.25}$$

The force per unit area on the lower plate ($x=0$) is then

$$\Pi^0_{xz} = \frac{\eta V}{h} - \frac{h}{2}\frac{\mathrm{d}p}{\mathrm{d}z} \tag{15.26}$$

and on the upper plate ($x=h$) becomes

$$\Pi^h_{xz} = \frac{\eta V}{h} + \frac{h}{2}\frac{\mathrm{d}p}{\mathrm{d}z}. \tag{15.27}$$

Obviously for simple Couette flow ($\mathrm{d}p/\mathrm{d}z=0$) the drag forces on both plates are the same,

$$\Pi^0_{xz} = \Pi^h_{xz} = \frac{\eta V}{h}. \tag{15.28}$$

15.2.3 Result for steady flow past a sphere (Stoke's formula)

If we assume $\mathrm{D}\boldsymbol{v}/\mathrm{D}t=0$, then the equations to be solved to determine the flow past a sphere are

$$\operatorname{div} \boldsymbol{v} = 0$$

and

$$\nabla^2 \boldsymbol{v} = \eta^{-1} \operatorname{grad} p.$$

These are often called Stoke's equations. For given boundary conditions, i.e. $\boldsymbol{v}=0$ at the surface of the sphere, Stoke's equations can be solved to determine $\boldsymbol{v}$. Once $\boldsymbol{v}$ is known, Π_{ij} at the surface of the sphere can be evaluated. The drag force experienced by the sphere has been found to be

$$\mathscr{F} = 6\pi\eta a V \tag{15.29}$$

where a is the radius of the sphere and V is the undisturbed flow velocity. Equation (15.29) is called Stoke's formula.

15.3 Reynolds number and turbulent flow

The nature of flow depends both on the geometry of the object as well as on physical quantities such as the density, viscosity, velocity, pressure, etc of the fluid. For instance, the flow pattern of water in a pipe is different from that in a river flowing past a rock; the flow in a given pipe is different for water and olive oil; slow- and fast-moving fluids etc. Obviously the mathematical description of the flow depends very much on the prevailing conditions. We saw earlier that for laminar flow in a capillary tube the mean flow velocity $v_0 \propto \Delta p$, provided that v_0 does not exceed a certain limit. Of course, it also depends upon the values of η and d (diameter of the tube). In other words, the flow in the tube remains laminar up to a certain condition and then it becomes turbulent. The question arises of whether one can prescribe the limiting condition as a single parameter which characterises the transition from laminar to tubulent flow.

The answer comes from the profound experimental and theoretical investigations of flow in tubes due to Reynolds. He examined the validity of the Hagen–Poiseuille formula or the transition from laminar to turbulent flow in tubes. The apparatus consisted of a glass tube fitted to a water reservoir at one end. By injecting a coloured fluid filament at the entrance of the tube, he observed its motion throughout the tube. For an average velocity less than a critical value the coloured filament remained in an undistorted path (i.e. laminar) throughout the flow. On the other hand, for velocities greater than the critical value the filament starts diffusing (i.e. turbulent) across the cross section of the tube. Reynolds in turn studied the effect of all possible variables, density (ρ), viscosity (η), flow velocity (v_0) and diameter of the tube (d) on the flow pattern. By making careful investigations he propounded that the only dimensionless number which could be framed out of these variables to characterise the flow pattern is

$$\mathscr{R} = \rho v_0 d/\eta. \tag{15.30}$$

$\mathscr{R}$ is called the Reynolds number. It is important to realise that in equation (15.30) the object, which is a tube here, is represented entirely by one parameter d. Therefore, use of (15.30) is not limited to flow in tubes, but rather it may also be used for other flows with suitable replacement for d. For instance, for flow past an object d may be replaced either by the length or by the diameter of the object.

When the Reynolds number of two different flows 1 and 2 are the same, i.e.

$$\frac{\rho_1 v_{01} d_1}{\eta_1} = \frac{\rho_2 v_{02} d_2}{\eta_2} \tag{15.31}$$

then the two flows are understood to be dynamically similar, i.e. both are either laminar or turbulent flows. This is known as the Reynolds law of similarity.

The values of Reynolds number vary over a wide range. It is of the order of 10^{-2} for blood plasma flowing in a blood vessel of $d \simeq 10^{-13}$ cm, $v_0 \simeq 2 \times 10^{-1}$ cm s^{-1}, $\eta \simeq 1.4 \times 10^{-2}$ P and $\rho \simeq 1.0$ g cm^{-3}; and $\mathscr{R} \sim 5 \times 10^7$ for an aircraft wing of length 3 m, $v_0 \simeq 900$ km h^{-1}, $\eta \simeq 1.8 \times 10^{-4}$ P and $\rho = 1.2$ kg m^{-3}.

So long as $\mathscr{R}$ is less than a critical value $\mathscr{R}_c$ the flow remains laminar irrespective of the individual values of ρ, v_0, η and d. When $\mathscr{R}$ exceeds $\mathscr{R}_c$, the transition from laminar to turbulent flow takes place. For the Hagen–Poiseuille flow, depending upon other conditions such as the inner surface of the tube, the shape of the inlet etc, the values of $\mathscr{R}_c$ vary approximately from 1000 to 20 000.

When the laminar flow in the tube becomes turbulent, i.e. $v_0 > v_{crit}$ then Δp (fall of pressure) is no longer in linear proportion to v_0, but rather it becomes

$$\Delta p \propto v_0^2. \tag{15.32}$$

This is known as the hydraulic law. A schematic variation of Δp and v_0 are shown in figure 15.4. Let us now attempt to obtain a pressure relation which is applicable to both laminar and turbulent flow in a tube. Since Δp and ρv_0^2 have the same dimensions, and also Δp is proportional to the length (l) of the tube, we express

$$\Delta p = (l\rho v_0^2/d)\mathscr{S} \tag{15.33}$$

where $\mathscr{S}$ must be a function of dimensionless quantities, i.e. $\mathscr{S} \to \mathscr{S}(l/d, \mathscr{R})$. If we take $\mathscr{S} = \beta\mathscr{R}^{-n}$, where β is a constant and n is a real variable, then (15.33) becomes

$$\Delta p = (\beta l \rho v_0^2/d)\mathscr{R}^{-n} \tag{15.34}$$

or

$$\Delta p \propto v_0^{2-n}. \tag{15.35}$$

For $n=1$, the linear relationship between Δp and v_0 of laminar flow is reproduced; and for $n=0$ the hydraulic law (15.32) is obtained. By making careful investigations it has however been found that n should vary between $\frac{1}{4}$ and $\frac{1}{5}$ for turbulent flow.

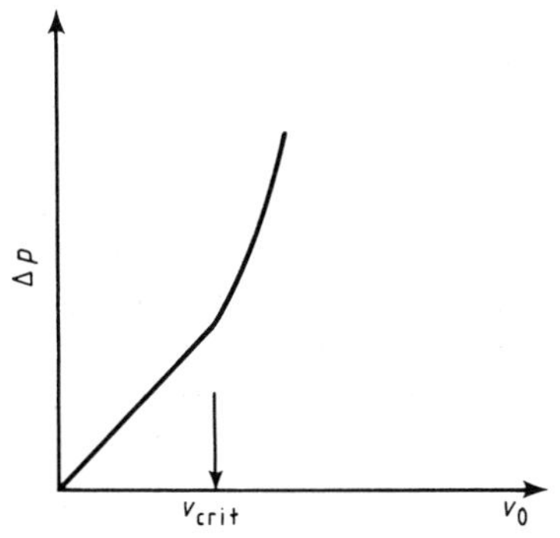

Figure 15.4 The pressure (Δp) in a tube as a function of the mean flow velocity (v_0). For $v_0 < v_{crit}$, the flow is laminar otherwise it is turbulent.

15.4 Potential flow on the basis of the NS equation in an incompressible fluid (Prandtl's boundary layer)

Earlier we spent two Chapters (11 and 12) discussing the potential flow for a perfect, i.e. inviscid, fluid on the basis of Euler's equation. Now we wish to apply the Navier–Stokes equation to discuss the potential flow problems in real fluids. If the fluid is incompressible (div $\mathbf{v}=0$) then the NS equation (13.16) for irrotational (curl $\mathbf{v}=0$) flow reduces to

$$\rho \frac{\partial \mathbf{v}}{\partial t}=\rho \boldsymbol{F}-\text{grad } p-\tfrac{1}{2}\rho \text{ grad } \mathbf{v}^2+\eta \nabla^2\mathbf{v}. \tag{15.36}$$

By virtue of the vector identity,

$$\nabla^2\mathbf{v}=\text{grad div } \mathbf{v}-\text{curl curl } \mathbf{v}=0 \tag{15.37}$$

equation (15.36) becomes

$$\rho \frac{\partial \mathbf{v}}{\partial t}=\rho \boldsymbol{F}-\text{grad } p-\tfrac{1}{2}\rho \text{ grad } \mathbf{v}^2. \tag{15.38}$$

It should be noted that the structure of equation (15.38) is the same as that of Euler's equation for potential flow. Thus all potential-flow solutions of Euler's equation are possible solutions of the NS equation. Since equation (15.38) is independent of the viscosity terms, it may appear at a glance that viscous forces have no effect on the solution of the equation if there is a potential flow. However this is not the case because they affect the solution through the boundary condition:

$$v_{\text{n}}=v_{\text{t}}=0 \qquad \text{for } \eta\neq 0. \tag{15.39}$$

On the other hand, for $\eta=0$ one simply has $v_{\text{n}}=0$. v_{n} and v_{t} are the normal and tangential components of the fluid velocity at the boundary of the solid object. Due to the extra boundary condition, i.e. $v_{\text{t}}=0$, the viscous forces influence the flow pattern in the region of the solid–fluid contact. For small η, the effect may be expected to last only for a short distance from the surface of the solid. The flow away from the solid surface remains unaffected as if the fluid has no viscosity. The boundary layer, or the region of influence of the viscosity around the solid object, is called Prandtl's boundary layer. The thickness, δ, of the boundary layer has been found to be proportional to $\eta^{1/2}$. In terms of Reynolds number, $\delta\propto\mathcal{R}^{-1/2}$. Without invoking the actual calculations performed by Prandtl, we provide here a simple argument which will lead us to Prandtl's observation. We recall from equation (14.38*a*) that the penetrating distance of the viscous or shear waves in a fluid is given by

$$\frac{1}{k_2}=\left(\frac{2\eta}{\rho_0\omega}\right)^{1/2}. \tag{15.40}$$

If we express $\omega \sim v/d$, d being some characteristic length, then (15.40) becomes

$$\frac{1}{k_2} \simeq d\left(\frac{2\eta}{\rho_0 v d}\right)^{1/2}$$
$$\simeq d\mathscr{R}^{-1/2}. \qquad (15.41)$$

$1/k_2$ which is the penetration length of the viscous waves, may be taken as Prandtl's boundary layer thickness δ and thus we obtain

$$\delta \propto \eta^{1/2} \qquad (15.42a)$$

or

$$\delta \propto \mathscr{R}^{-1/2}. \qquad (15.42b)$$

Problems

15.1 Show that the total drag force ($\mathscr{F}$) acting on the tube in the Hagen–Poiseuille flow is given by

$$\mathscr{F} = -(\pi a^2 \, \Delta p) \qquad (P15.1)$$

where a is the radius of the tube and $\Delta p = p_1 - p_2$ (p_1 and p_2 are the end pressures).

15.2 Verify equation (15.29).

15.3 A viscous fluid flows with a steady velocity $v(z)$ along the x-axis between two fixed plane parallel plates at $z=0$ and $z=h$. Assume that the flow is laminar and that the planes are sufficiently wide in the y-direction that the problem can be treated in two dimensions. Show that

(i) $\mathrm{d}p/\mathrm{d}x$ is a constant

(ii)
$$v(z) = \frac{1}{2\eta}\frac{\mathrm{d}p}{\mathrm{d}x}\, z(z-h) \qquad (P15.2)$$

(iii) What is the tangential force on the two plates per unit area?

APPENDIX A

PROOF THAT STRESS IS A TENSOR

The proof that σ_{ij} is a tensor follows by considering the force vector due to it. Let us consider a tetrahedron-shaped element of the body OABC which is in equilibrium. The three faces of the tetrahedron, i.e. OBC, OAC and OAB are the coordinate planes $x_1=0$, $x_2=0$ and $x_3=0$ respectively. The stress components σ_{ij} acting on each face are shown in figure A.1. The fourth face ABC here represents a variable element of surface which is perpendicular to a unit vector $\boldsymbol{n}$. If $\mathrm{d}A$ is the area of the face ABC, $\mathrm{d}A_1$ of BOC, $\mathrm{d}A_2$ of AOC and $\mathrm{d}A_3$ of AOB; the equilibrium condition for the force in the x_1-direction is then

$$\mathscr{F}_1 \mathrm{d}A = \sigma_{11}\, \mathrm{d}A_1 + \sigma_{21}\, \mathrm{d}A_2 + \sigma_{31}\, \mathrm{d}A_3. \tag{A.1}$$

Since the areas $\mathrm{d}A_1$, $\mathrm{d}A_2$ and $\mathrm{d}A_3$ may be considered as the projections of $\mathrm{d}A$, i.e.

$$\mathrm{d}A_1/\mathrm{d}A = \cos(\boldsymbol{n}, x_1) \equiv n_1 \tag{A.2a}$$

$$\mathrm{d}A_2/\mathrm{d}A = \cos(\boldsymbol{n}, x_2) \equiv n_2 \tag{A.2b}$$

$$\mathrm{d}A_3/\mathrm{d}A = \cos(\boldsymbol{n}, x_3) \equiv n_3 \tag{A.2c}$$

equation (A.1) becomes

$$\mathscr{F}_1 = n_1\sigma_{11} + n_2\sigma_{21} + n_3\sigma_{31}. \tag{A.3a}$$

Similarly, for the other two directions one has

$$\mathscr{F}_2 = n_1\sigma_{12} + n_2\sigma_{22} + n_3\sigma_{32} \tag{A.3b}$$

$$\mathscr{F}_3 = n_1\sigma_{13} + n_2\sigma_{23} + n_3\sigma_{33} \tag{A.3c}$$

or in short,

$$\mathscr{F}_i = n_j\sigma_{ji}. \tag{A.4}$$

Since $\mathscr{F}$ and $\boldsymbol{n}$ are vectors, σ must be a tensor (see properties of tensors in §2.4).

It may also be shown that σ_{ij} conform to the transformation rule of second-rank tensors and therefore it should be regarded as a tensor of rank two. Actually $\mathscr{F}_1$, $\mathscr{F}_2$ and $\mathscr{F}_3$ represent the components of stress on the n-surface acting along the x_1-, x_2- and x_3-directions respectively, i.e.

$$\mathscr{F}_1 = \sigma_{n1} \qquad \mathscr{F}_2 = \sigma_{n2} \qquad \text{and } \mathscr{F}_3 = \sigma_{n3} \tag{A.5}$$

therefore,

$$\sigma_{nn} = n_1\mathscr{F}_1 + n_2\mathscr{F}_2 + n_3\mathscr{F}_3 \tag{A.6}$$

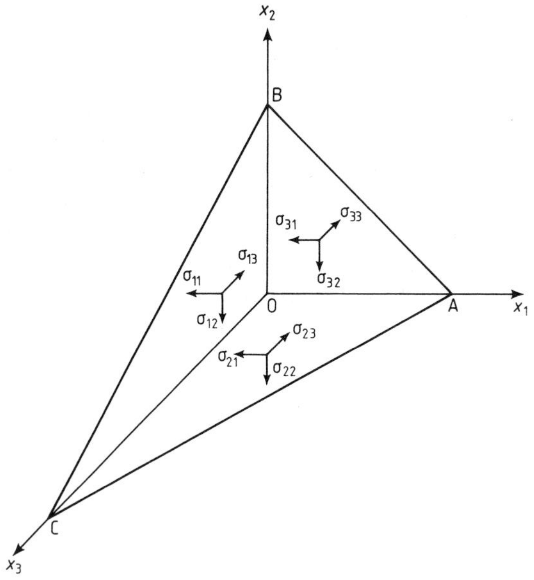

Figure A.1 The stress components on the faces of a tetrahedron-shaped body.

Now let us denote ***n*** as the primed axis x'_1 and the other two axes in the plane as x'_2 and x'_3 then following the transformation rule of table 2.1, one writes

$$\sigma'_{11}=a_{11}\mathscr{F}_1+a_{12}\mathscr{F}_2+a_{13}\mathscr{F}_3 \tag{A.7a}$$

$$\sigma'_{12}=a_{21}\mathscr{F}_1+a_{22}\mathscr{F}_2+a_{23}\mathscr{F}_3 \tag{A.7b}$$

$$\sigma'_{13}=a_{31}\mathscr{F}_1+a_{32}\mathscr{F}_2+a_{33}\mathscr{F}_3 \tag{A.7c}$$

or in abbreviated form,

$$\sigma'_{ij}=a_{jl}\mathscr{F}_l. \tag{A.8}$$

On the other hand, we have

$$\mathscr{F}_1=a_{11}\sigma_{11}+a_{12}\sigma_{21}+a_{13}\sigma_{31} \tag{A.9a}$$

$$\mathscr{F}_2=a_{11}\sigma_{12}+a_{12}\sigma_{22}+a_{13}\sigma_{32} \tag{A.9b}$$

$$\mathscr{F}_3=a_{11}\sigma_{13}+a_{12}\sigma_{23}+a_{13}\sigma_{33} \tag{A.9c}$$

or,

$$\mathscr{F}_l=a_{1k}\sigma_{kl}. \tag{A.10}$$

From equations (A.8) and (A.10) one therefore has

$$\sigma'_{1j}=a_{1k}a_{jl}\sigma_{kl}. \tag{A.11}$$

More generally we express

$$\sigma'_{ij}=a_{ik}a_{jl}\sigma_{kl}. \tag{A.12}$$

Thus σ_{ij} transform like components of a second-rank tensor. This is our final result, and it should be emphasised that it is independent of the size of the tetrahedron. It is therefore true for any arbitrary n-plane passing through a point O. Equation (A.12) also holds even if the tetrahedron is subjected to the body forces, i.e. for the tetrahedron being in a non-equilibrium state. The extra term appearing due to body forces which is a function of the volume of the tetrahedron vanishes in the limit that the tetrahedron reduces to a point.

APPENDIX B

ATTENUATION OF SOUND ENERGY DUE TO VISCOSITY

Equation (14.25) for the attenuation of sound waves can also be derived from the expression (13.22) for the energy dissipation due to viscosity. For the one-dimensional case we recall that the amount of energy lost per unit volume per unit time is given by

$$\left(\frac{\delta \varepsilon}{\delta t}\right)_{\text{vis}} = (b + \tfrac{4}{3}\eta)\left(\frac{\partial v_1}{\partial x_1}\right)^2. \tag{B.1}$$

For a plane progressive wave v_1 varies sinusoidally,

$$v_1 = v_1^0 \cos(\omega t - k_1 x_1) \tag{B.2}$$

therefore we take the average of (B.1), i.e.

$$\Delta E = \overline{\left(\frac{\delta \varepsilon}{\delta t}\right)}_{\text{vis}} = (b + \tfrac{4}{3}\eta)\overline{\left(\frac{\partial v_1}{\partial x_1}\right)}^2. \tag{B.3}$$

The total energy of a plane progressive wave, on the other hand, consists of equal amounts of kinetic and potential energies. The total energy per unit volume may be expressed as

$$\begin{aligned} E &= \tfrac{1}{2}\rho_0 \overline{v_1^2} + \tfrac{1}{2}\rho_0 \overline{v_1^2} \\ &= \rho_0 \overline{v_1^2}. \end{aligned} \tag{B.4}$$

Therefore, the fractional loss in energy per unit time may be expressed as

$$\begin{aligned} \frac{\Delta E}{E} &= (b + \tfrac{4}{3}\eta)\overline{\left(\frac{\partial v_1}{\partial x_1}\right)}^2 (\rho_0 \overline{v_1^2})^{-1} \\ &\equiv \frac{(b + \tfrac{4}{3}\eta) k_1^2 (v_1^0)^2 \overline{\sin^2(\omega t - k_1 x_1)}}{\rho (v^0)^2 \overline{\cos^2(\omega t - k_1 x_1)}} \\ &= (b + \tfrac{4}{3}\eta) k_1^2 \rho_0^{-1}. \end{aligned} \tag{B.5}$$

Since the wave travels a distance of C_0 cm s^{-1}, the fractional loss in energy per cm is

$$= (b + \tfrac{4}{3}\eta) k_1^2 (\rho_0 C_0)^{-1}. \tag{B.6}$$

From the definition of the amplitude attenuation the above relation must be equal to $2k_2$, i.e.

$$\begin{aligned} k_2 &= (b+\tfrac{4}{3}\eta)k_1^2(2\rho_0 C_0)^{-1} \\ &= (b+\tfrac{4}{3}\eta)\omega^2(2\rho_0 C_0^3)^{-1}. \end{aligned} \tag{B.7}$$

(B.7) is the same as equation (14.25*a*).

BIBLIOGRAPHY

A: Text books of immediate interest

As we mentioned earlier no single elementary text available in the area contains all the topics of the present book. The following books, however, cover many of the topics.

Fetter L A and Walecka J D 1980 *Theoretical Mechanics of Particles and Continua* (New York: McGraw–Hill)
This book provides a good account of the growth of the theory of continuous media starting from particle mechanics. The discussion of elasticity and potential flow is brief but chapters like 9 to 13 are of immediate interest and have a large number of suitable problems which are set.

Fung Y C 1977 *A First Course in Continuum Mechanics* 2nd edn (New Jersey: Prentice–Hall)
This elementary text is largely devoted to solids and contains a large number of problems. Readers may find it useful for first-hand reading.

Paterson A R 1983 *A First Course in Fluid Mechanics* (Cambridge: Cambridge University Press)
This book covers many of our topics in fluid mechanics.

Sommerfeld A 1950 *Mechanics of Deformable Bodies* (New York: Academic)
Old but still we find it the most useful book available in the area. It is brief on sound propagation but good for many other topics. Readers should realise that this work has had a great influence on the structure of the present book.

Whitaker S 1968 *Introduction to Fluid Mechanics* (New Jersey: Prentice–Hall)
Readers may find this book quite useful as supplementary reading for the Chapters on fluids.

B: Standard reference books

Bhatia A B 1967 *Ultrasonic Absorption* (Oxford: Clarendon)
Though this book is exclusively devoted to a detailed treatment of ultrasonic absorption in material media, our readers may particular enjoy reading the Chapters on the propagation of elastic waves in solids and liquids. We recommend Chapters 2 to 4, 11 and 12 for immediate interest.

Lamb H 1945 *Hydrodynamics* 6th edn (New York: Dover)
This is a well known treatise on hydrodynamics which is famous for its complete mathematical account of the theory of the motion of fluids. Chapters 1 to 7, 10 and 11 are directly relevant.

Landau L D and Lifshitz E M 1959a *Theory of Elasticity* (London: Pergamon)
We recommend this book for further reading about elasticity

—— 1959b *Fluid Mechanics* (London: Pergamon)
Chapters 1–2, 5 and 8 are recommended for further reading.

Love A E H 1944 *A Treatise on the Mathematical Theory of Elasticity* 4th edn (New York: Dover)
This is a vast book on the theory of elasticity; however, the first few Chapters on stress and strain are very useful from the viewpoint of a first reading.

Milne-Thomson L M 1960 *Theoretical Hydrodynamics* (London: Macmillan)
This book contains a huge literature on theoretical hydrodynamics and is very good for quick reference.

Nye J F 1957 *Physical Properties of Crystals* (Oxford: Clarendon)
Part 1 of this book is recommended for supplementary reading for our Chapter 2. In addition to the Chapters on elasticity, this book provides a good account of the theoretical formulation of the electrical and magnetic properties of solids.

Ramsey A S 1935 *A Treatise on Hydrodynamics, Part II. Hydrodynamics* 4th edn (London: Bell and Sons)
Obviously this is again a very old book but covers many of our topics on fluids and much more. Our work on fluids has greatly benefited from this monumental work.

C: General references

Aitken A C 1956 *Determinants and Matrices* 9th edn (Edinburgh: Oliver and Boyd)
Arenberg D L 1948 *J. Acoust. Soc. Am.* **20** 1
Bhagavantam S 1966 *Crystal Symmetry and Physical Properties* (London: Academic)
Cottrell A H 1953 *Dislocations and Elastic Flow in Crystals* (Oxford: Clarendon)
Goldstein H 1980 *Classical Mechanics* 2nd edn (New York: Addison–Wesley)
Green A E and Adkins J E 1970 *Large Elastic Deformation* (Oxford: Clarendon)
Green A E and Zerna W 1954 *Theoretical Elasticity* (Oxford: Clarendon)
Haymes R C 1971 *Introduction to Space Science* (New York: Wiley) p 56
Hirth J P and Lothe J 1968 *Theory of Dislocations* (New York: McGraw–Hill)
Kolsky H 1953 *Stress Waves in Solids* (Oxford: Clarendon)
Koster W and Franz H 1961 *Metall. Rev.* **6** 1
Litovitz T A 1960 *Non-crystalline Solids* ed V D Frechette (New York: Wiley) p 252
Litovitz T A and Sette D 1953 *J. Chem. Phys.* **25** 17
McLellan A G 1980 *The Classical Thermodynamics of Deformable Materials* (Cambridge: Cambridge University Press)
Mason P Warren 1958 *Physical Acoustics and the Properties of Solids* (Princeton, NJ: van Nostrand)
Mathews J and Walker R L 1970 *Mathematical Methods of Physics* 2nd edn (Menlo Pk, California: W A Benjamin)
Murnaghan F D 1951 *Finite Deformation of an Elastic Solid* (New York: Wiley)
Musgrave N J P 1970 *Crystal Acoustics* (San Francisco: Holden-Day)
Nabarro F R N 1967 *Theory of Crystal Dislocations* (Oxford: Clarendon)
Phillips F C 1971 *An Introduction to Crystallography* 4th edn (Edinburgh: Oliver and Boyd)
Read W T 1953 *Dislocations in Crystals* (New York: McGraw–Hill)
Simmons G and Wang H 1971 *Single Crystal Elastic Constants and Calculated Aggregate Properties, a Handbook* (Cambridge, Massachusetts: MIT Press)
Stanley W 1984 *Rev. Mod. Phys.* **56** 41
Voigt W 1910 *Lehrbuch der Kristallphysik* (Leipzig: Teubner)
Wallace D C 1972 *Thermodynamics of Crystals* (New York: Wiley)
Weast R C (ed.) 1984 *CRC Handbook of Chemistry and Physics* (Boca Raton, Florida: Chemical Rubber)
Wooster W A 1973 *Tensors and Group Theory for the Physical Properties of Crystals* (Oxford: Clarendon)
Zemansky M W 1968 *Heat and Thermodynamics* 5th edn (New York: McGraw–Hill)

INDEX